Basic Earthquake Engineering

Basic Earthquake Engineering

Editor

Petre Kozel

scitus
academics

Basic Earthquake Engineering

Edited by **Petre Kozel**

Printed in 2017

ISBN: 978-1-68117-110-4
Library of Congress Control Number: 2015950360

© 2016 by
SCITUS Academics LLC,
616, Corporate Way, Suite 2, 4766,
Valley Cottage, NY 10989

www.scitusacademics.com

Table of Content

Preface

Earthquake engineering or seismic engineering is the scientific field concerned with protecting society, the natural environment, and the man-made environment from earthquakes by limiting the seismic risk to socio-economically acceptable levels. Earthquake Engineering can be defined as the branch of engineering devoted to mitigating earthquake hazards. In this broad sense, earthquake engineering covers the investigation and solution of the problems created by damaging earthquakes, and consequently the work involved in the practical application of these solutions, i.e. in planning, designing, constructing and managing earthquake-resistant structures and facilities.

The main objectives of earthquake engineering are to predict the potential consequences of strong earthquakes on urban areas and civil infrastructure. A properly engineered structure does not necessarily have to be extremely strong or expensive. It has to be properly designed to withstand the seismic effects while sustaining an acceptable level of damage.

This book emphasize to students of structural and architectural engineering the problems and solutions in attaining efficient earthquake-resistant structures and facilities. To achieve this objective, after a brief discussion of the general goals in seismic-resistant design and construction of structures and facilities, the diverse sources of damage that can be triggered by an earthquake are discussed.

CHAPTER 1
A Cognitive Look at Geotechnical Earthquake Engineering: Understanding The Multidimensionality of The Phenomena

Silvia Garcia

Geotechnical Department, Institute of Engineering, National University of Mexico, Mexico

INTRODUCTION

Essentially, disasters are human-made. For a catastrophic event, whether precipitated by natural phenomena or human activities, assumes the state of a disaster when the community or society affected fails to cope. Earthquake hazards themselves do not necessarily lead to disasters, however intense, inevitable or unpredictable, translate to disasters only to the extent that the population is unprepared to respond, unable to deal with, and, consequently, severely affected. Seismic disasters could, in fact, be reduced if not prevented. With today's advancements in science and technology, including early warning and forecasting of the natural phenomena, together with innovative approaches and strategies for enhancing local capacities, the impact of earthquake hazards somehow could be predicted and mitigated, its detrimental effects on populations reduced, and the communities adequately protected.

After each major earthquake, it has been concluded that the experienced ground motions were not expected and soil behavior and soil-structure interaction were not properly predicted. Failures, associated to inadequate design/construction and to lack of phenomena comprehension, obligate further code reinforcement and research. This scenario will be repeated after each earthquake. To overcome this issue, *Earthquake Engineering* should change its views on the present methodologies and techniques toward more scientific, doable, affordable, robust and adaptable solutions.

A competent modeling of engineering systems, when they are affected by seismic activity, poses many difficult challenges. Any representation designed for reasoning about models of such systems has to be flexible enough to handle various degrees of complexity and uncertainty, and at the same time be sufficiently powerful to deal with situations in which the input signal may or may not be controllable. Mathematically-based models are developed using scientific theories and concepts that just apply to particular conditions. Thus, the core of the model comes from assumptions that for complex systems usually lead to simplifications (perhaps oversimplifications) of the problem phenomena. It is fair to argue that the representativeness of a particular theoretical model largely depends on the degree of comprehension the developer has on the behavior of the actual engineering problem. Predicting natural-phenomena characteristics like those of earthquakes, and thereupon their potential effects at particular sites, certainly belong to a class of problems we do not fully understand. Accordingly, analytical modeling often becomes the bottleneck in the development of more accurate procedures. As a consequence, a strong demand for advanced modeling an identification schemes arises.

Cognitive Computing CC technologies have provided us with a unique opportunity to establish coherent seismic analysis environments in which uncertainty and partial data-knowledge are systematically handled. By seamlessly combining learning, adaptation, evolution, and fuzziness, CC complements current engineering approaches allowing us develop a more comprehensive and unified framework to the effective management of earthquake phenomena. Each CC algorithm has well-defined labels and could usually be identified with specific scientific communities. Lately, as we improved our understanding of these algorithms' strengths and weaknesses, we began to leverage their best features and developed hybrid algorithms that indicate a new trend of co-existence and integration between many scientific communities to solve a specific task.

In this chapter geotechnical aspects of earthquake engineering under a cognitive examination are covered. Geotechnical earthquake engineering, an area that deals with the design and construction of projects in order to resist the effect of earthquakes, requires an understanding of geology, seismology and earthquake engineering. Furthermore, practice of geotechnical earthquake engineering also requires consideration of social, economic and political factors. Via the development of cognitive interpretations of selected topics: i) spatial variation of soil dynamic properties, ii) attenuation laws for rock sites (seismic input), iii) generation of artificial-motion time histories, iv) effects of local site conditions (site effects), and iv) evaluation of liquefaction susceptibility, CC techniques (Neural Networks NNs, Fuzzy Logic FL and Genetic Algorithms GAs) are presented as appealing alternatives for integrated data-driven and theoretical procedures to generate reliable seismic models.

GEOTECHNICAL EARTHQUAKE HAZARDS

The author is well aware that standards for geotechnical seismic design are under development worldwide. While there is no need to "reinvent the wheel" there is a requirement to adapt such initiatives to fit the emerging safety philosophy and demands. This investigation also strongly endorses the view that "guidelines" are far more desirable than "codes" or "standards" disseminated all over seismic regions. Flexibility in approach is a key ingredient of geotechnical engineering and the cognitive technology in this area is rapidly advancing. The science and practice of geotechnical earthquake engineering is far from mature and need to be expanded and revised periodically in coming years. It is important that readers and users of the computational models presented here familiarize themselves with the latest advances and amend the recommendations herein appropriately.

This document is not intended to be a detailed treatise of latest research in geotechnical earthquake engineering, but to provide sound guidelines to support rational cognitive approaches. While every effort has been made to make the material useful in a wider range of applications, applicability of the material is a matter for the user to judge. The main aim of this guidance document is to promote consistency of cognitive approach to everyday situations and, thus, improve geotechnical-earthquake aspects of the performance of the built safe-environment.

A "Soft" Interpretation of Ground Motions

After a sudden rupture of the earth's crust (caused by accumulating stresses, elastic strain-energy) a certain amount of energy radiates from the rupture as seismic waves. These waves are attenuated, refracted, and reflected as they travel through the earth, eventually reaching the surface where they cause ground shaking. The principal geotechnical hazards associated with this event are fault rupture, ground shaking, liquefaction and lateral spreading, and landsliding. Ground shaking is one of the principal seismic hazards that causes extensive damage to the built environment and failure of engineering systems over large areas. Earthquake loads and their effects on structures are directly related to the intensity and duration of ground shaking. Similarly, the level of ground deformation, damage to earth structures and ground failures are closely related to the severity of ground shaking.

In engineering evaluations, three characteristics of ground shaking are typically considered: i) the amplitude, ii) frequency content and iii) significant duration of shaking (time over which the ground motion has relatively significant amplitudes).These characteristics of the ground motion at a given site are affected by numerous complex factors such as the source mechanism, earthquake magnitude, rupture directivity, propagation path of seismic waves, source distance and effects of local soil conditions. There are many unknowns and uncertainties associated with these issues which in turn result in significant uncertainties regarding the characteristics of the ground motion and earthquake loads.

If the random nature of response to earthquakes (aleatory uncertainty) cannot be avoided [1,2], it is our limited knowledge about the patterns between seismic events and their manifestations -ground motions- at a site (epistemic uncertainty) that must be improved thorough more scientific seismic analyses. A strategic factor in seismic hazard analysis is the ground motion model or attenuation relation. These attenuation relationships has been developed based on magnitude, distance and site category, however, there is a tendency to incorporate other parameters, which are now known to be significant, as the tectonic environment, style of faulting and the effects of topography, deep basin edges and rupture directivity. These distinctions are recognized in North America, Japan and New Zealand [3-6], but ignored in most other regions of the world [7]. Despite recorded data suggest that ground motions depend, in a significant way, on these aspects, these inclusions did not have had a remarkable effect on the predictions confidence and the geotechnical earthquake engineer prefers the basic and clear-cut approximations on those that demand a *blind* use of coefficients or an intricate determination of soil/fault conditions.

A key practice in current aseismic design is to develop design spectrum compatible time histories. This development entails the modification of a time history so that its response spectrum matches within a prescribed tolerance level, the target design spectrum. In such matching it is important to retain the phase characteristics of the selected ground motion time history. Many of the techniques used to develop compatible motions do not retain the phase [8]. The response spectrum alone does not adequately characterize specific-fault ground motion. Near-fault ground motions must be characterized by a long period pulse of strong motion of a fairly brief duration rather than the stochastic process of long duration that characterizes more distant ground motions. Spectrum compatible with these specific motions will not have these characteristics unless the basic motion being modified to ensure compatibility has these effects included. Spectral compatible motions could match the entire spectrum but the problem arises on finding a "real" earthquake time series that match the specific nature of ground motion. For nonlinear analysis of structures, spectrum compatible motions should also correspond to the particular energy input [9], for this reason, designers should be cautious about using spectrum compatible motions when estimating the displacements of embankment dams and earth structures under strong shaking, if the acceptable performance of these structures is specified by criteria based on tolerable displacements.

Another important seismic phenomenon is the liquefaction. Liquefaction is associated with significant loss of stiffness and strength in the shaken soil and consequent large ground deformation. Particularly damaging for engineering structures are cyclic ground movements during the period of shaking and excessive residual deformations such as settlements of the ground and lateral spreads. Ground surface disruption including surface cracking, dislocation, ground distortion, slumping and permanent deformations, large settlements and lateral spreads are commonly observed at liquefied sites. In sloping ground and backfills behind retaining structures in waterfront areas, liquefaction often results in large permanent ground displacements in the down-slope direction or towards waterways (lateral spreads). Dams, embankments and sloping ground near riverbanks where certain shear strength is required for stability under gravity loads are particularly prone to such failures. Clay soils may also suffer some loss of strength during shaking but are not subject to boils and other "classic" liquefaction phenomena. For intermediate soils, the transition from "sand like" to "clay-like" behavior depends primarily on whether the soil is a matrix of coarse grains with fines contained within the pores or a matrix of plastic fines with coarse grained "filler". Recent papers by Boulanger and Idriss [10, 11] are helpful in clarifying issues surrounding the liquefaction and strain softening of different soil types during strong ground shaking. Engineering judgment based on good quality investigations and data interpretation should be used for classifying such soils as liquefiable or non-liquefiable.

Procedures for evaluating liquefaction, potential and induced lateral spread, have been studied by many engineering committees around the world. The objective has been to review research and field experience on liquefaction and recommended standards for practice. Youd and Idriss [12] findings and the liquefaction-resistance chart proposed by Seed et al. [13] in 1985, stay as standards for practice. They have been slightly modified to adjust new registered input-output conditions and there is a strong tendency to recommend i) the adoption of the cone penetration test CPT, standard penetration test SPT or the shear wave velocities for describing the *in situ* soil conditions [14] and ii) the modification of magnitude factors used to convert the critical stress ratios from the liquefaction assessment charts (usually developed for M7:5) to those appropriate for earthquakes of diverse magnitudes [12, 15].

COGNITIVE COMPUTING

Cognitive Computing CC as a discipline in a narrow sense is an application of computers to solve a given computational problem by imperative instructions; while in a broad sense, it is a process to implement the instructive intelligence by a system that transfers a set of given information or instructions into expected behaviors. According to theories of cognitive informatics [16-18], computing technologies and systems may be classified into the categories of imperative, autonomic, and cognitive from the bottom up. Imperative computing is a traditional and passive technology based on stored-program controlled behaviors for data processing [19-24]. An autonomic computing is goal-driven and self-decision-driven technologies that do not rely on instructive and procedural information [25-28]. Cognitive computing is more intelligent technologies beyond imperative and autonomic computing, which embodies major natural intelligence behaviors of the brain such as thinking, inference, learning, and perceptions.

Cognitive computing is an emerging paradigm of intelligent computing methodologies and systems, which implements computational intelligence by autonomous inferences and perceptions mimicking the mechanisms of the brain. This section presents a brief description on the theoretical framework and architectural techniques of cognitive computing beyond conventional imperative and autonomic computing technologies. Cognitive models are explored on the basis of the latest advances in applying computational intelligence. These applications of cognitive computing are described from the aspects of cognitive search engines, which demonstrate how machine and computational intelligence technologies can drive us toward autonomous knowledge processing.

Computational Intelligence: Soft Computing Technologies

The *computational intelligence* is a synergistic integration of essentially three computing paradigms, viz. neural networks, fuzzy logic and evolutionary computation entailing probabilistic reasoning (belief networks, genetic algorithms and chaotic systems) [29]. This synergism provides a framework for flexible information processing applications designed to operate in the real world and is commonly called *Soft Computing SC* [30]. Soft computing technologies are robust by design, and operate by trading off precision for tractability. Since they can handle uncertainty with ease, they conform better to real world situations and provide lower cost solutions.

The three components of soft computing differ from one another in more than one way. Neural networks operate in a numeric framework, and are well known for their learning and generalization capabilities. Fuzzy systems [31] operate in a linguistic framework, and their strength lies in their capability to handle linguistic information and perform approximate reasoning. The evolutionary computation techniques provide powerful search and optimization methodologies. All the three facets of soft computing differ from one another in their time scales of operation and in the extent to which they embed *a priori* knowledge.

Figure 1 shows a general structure of Soft Computing technology. The following main components of SC are known by now: fuzzy logic FL, neural networks NN, probabilistic reasoning PR, genetic algorithms GA, and chaos theory ChT (Figure 1). In SC FL is mainly concerned with imprecision and approximate reasoning, NN with learning, PR with uncertainty and propagation of belief, GA with global optimization and search and ChT with nonlinear dynamics. Each of these computational paradigms (emerging reasoning technologies) provides us with complementary reasoning and searching methods to solve complex, real-world problems. In large scope, FL, NN, PR, and GA are complementary rather that competitive [32-34]. The interrelations between the components of SC, shown in Figure 1, make the theoretical foundation of Hybrid Intelligent Systems. As noted by L. Zadeh: "… the term hybrid intelligent systems is gaining currency as a descriptor of systems in which FL, NC, and PR are used in combination. In my view, hybrid intelligent systems are the wave of the future" [35]. The use of Hybrid Intelligent Systems are leading to the development of numerous manufacturing system, multimedia system, intelligent robots, trading systems, which exhibits a high level of MIQ (machine intelligence quotient).

Comparative Characteristics of SC Tools

The constituents of SC can be used independently (fuzzy computing, neural computing, evolutionary computing etc.), and more often in combination [36, 37, 38- 40, 41]. Based on independent use of the constituents of Soft Computing, fuzzy technology, neural technology, chaos technology and others have been recently applied as emerging technologies to both industrial and non-industrial areas.

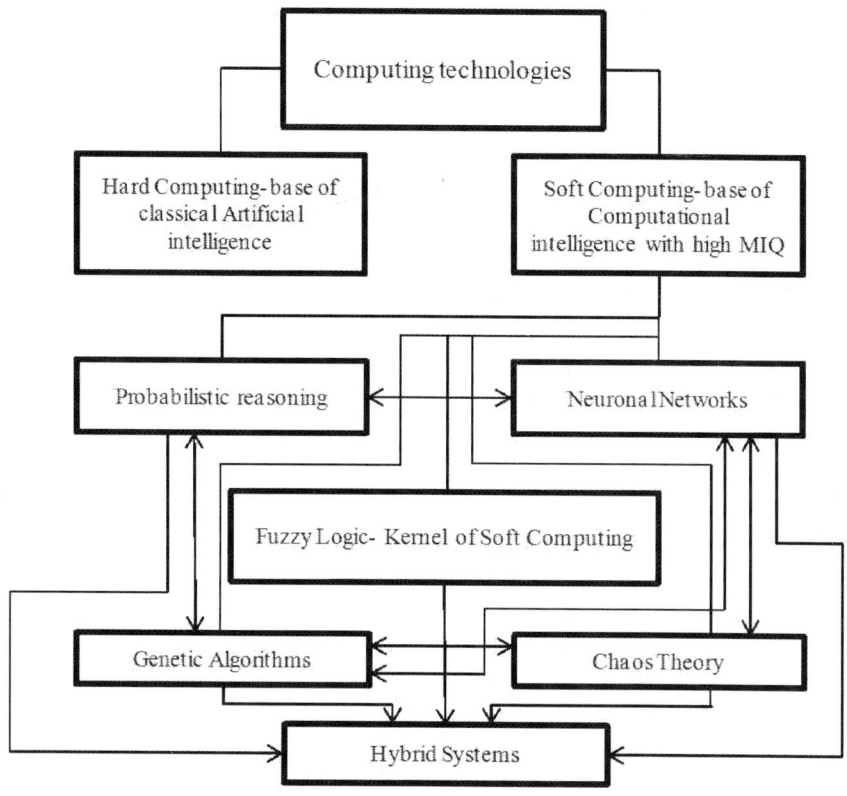

Figure 1. Soft Computing Components.

Fuzzy logic is the leading constituent of Soft Computing. In Soft Computing, fuzzy logic plays a unique role. FL serves to provide a methodology for computing [36]. It has been successfully applied to many industrial spheres, robotics, complex decision making and diagnosis, data compression, and many other areas. To design a system processor for handling knowledge represented in a linguistic or uncertain numerical form we need a fuzzy model of the system. Fuzzy sets can be used as a universal approximator, which is very important for modeling unknown objects. If an operator cannot tell linguistically what kind of action he or she takes in a specific situation, then it is quite useful to model his/her control actions using numerical data. However, fuzzy logic in its so called *pure form* is not always useful for easily constructing intelligent systems. For example, when a designer does not have sufficient prior information (knowledge) about the system, development of acceptable fuzzy rule base becomes impossible. As the complexity of the system increases, it becomes difficult to specify a correct set of rules and membership functions for describing adequately the behavior of the system. Fuzzy systems also have the disadvantage of not being able to extract additional knowledge from the experience and correcting the fuzzy rules for improving the performance of the system.

Another important component of Soft Computing is neural networks. Neural networks NN viewed as parallel computational models are parallel fine-grained implementation of non-linear static or dynamic systems. A very important feature of these networks is their adaptive nature, where "learning by example" replaces traditional "programming" in problems solving. Another key feature is the intrinsic parallelism that allows fast computations. Neural networks are viable computational models for a wide variety of problems including pattern classification, speech synthesis and recognition, curve fitting, approximation capability, image data compression, associative memory, and modeling and control of non-linear unknown systems [42, 43]. NN are favorably distinguished for efficiency of their computations and hardware implementations. Another advantage of NN is generalization ability, which is the ability to classify correctly new patterns. A significant disadvantage of NN is their poor interpretability. One of the main criticisms addressed to neural networks concerns their black box nature [35].

Evolutionary Computing EC is a revolutionary approach to optimization. One part of EC—genetic algorithms—are algorithms for global optimization. Genetic algorithms GAs are based on the mechanisms of natural selection and genetics [44]. One advantage of genetic algorithms is that they effectively implement parallel multi-criteria search. The mechanism of genetic algorithms is simple. Simplicity of operations and powerful computational effect are the two main advantages of genetic algorithms. The disadvantages are the problem of convergence and the absence of strong theoretical foundation. The requirement of coding the domain of the real variables' into bit strings also seems to be a drawback of genetic algorithms. It should be also noted that the computational speed of genetic algorithms is low.

Because in this investigation PR and ChT are not exploited, they are not going to be explained. For the interested reader [41] is recommended. Table 1 presents the comparative characteristics of the components of Soft Computing. For each component of Soft Computing there is a specific class of problems, where the use of other components is inadequate.

Intelligent Combinations of SC

As it was shown above, the components of SC complement each other, rather than compete. It becomes clear that FL, NC and GA are more effective when used in combinations. Lack of interpretability of neural networks and poor learning capability of fuzzy systems are similar problems that limit the application of these tools. Neurofuzzy systems are hybrid systems which try to solve this problem by combining the learning capability of connectionist models with the interpretability property of fuzzy systems. As it was noted above, in case of dynamic work environment, the automatic knowledge base correction in fuzzy systems becomes necessary. On the other hand, artificial neural networks are successfully used in problems connected to knowledge acquisition using learning by examples with the required degree of precision.

Incorporating neural networks in fuzzy systems for fuzzification, construction of fuzzy rules, optimization and adaptation of fuzzy knowledge base and implementation of fuzzy reasoning is the essence of the Neurofuzzy approach.

Table 1: Central characteristics of Soft Computing technologies

	Fuzzy Sets	Artificial Neural Networks	Evolutionary Computing, GA	Probabilistic Reasoning	Chaotic computing
Weaknesses	•Knowledge acquisition •Learning	•Black Box interpretability	•Coding •Computational speed	•Limitation of the axioms of Probability Theory •Lack of complete knowledge •Copmputational complexity	•Computational complexity •Chaos identification complexity
Strengths	•Interpretability •Transparency •Plausibility •Graduality •Modeling •Reasoning •Tolerance to imprecision	•Learning •Adaptation •Fault tolerance •Curve fitting •Generalization ability •Approximation ability	Computational efficiency •Global optimization	•Rigorous framework •Well understanding	•Nonlinear dynamics simulation •Discovering chaos in observed data (with noise) •Determinig the predictability •Prediction strategies formulation

The combination of genetic algorithms with neural networks yields promising results as well. It is known that one of main problems in development of artificial neural systems is selection of a suitable learning method for tuning the parameters of a neural network (weights, thresholds, and structure). The most known algorithm is the "error back propagation" algorithm. Unfortunately, there are some difficulties with "back propagation". First, the effectiveness of the learning considerably depends on initial set of weights, which are generated randomly. Second, the "back propagation", like any other gradient-based method, does not avoid local minima. Third, if the learning rate is too slow, it requires too much time to find the solution. If, on the other hand, the learning rate is too high it can generate oscillations around the desired point in the weight space. Fourth, "back propagation" requires the activation functions to be differentiable. This condition does not hold for many types of neural networks. Genetic algorithms used for solving many optimization problems when the "strong" methods fail to find appropriate solution, can be successfully applied for learning neural networks, because they are free of the above drawbacks.

The models of artificial neurons, which use linear, threshold, sigmoidal and other transfer functions are effective for neural computing. However, it should be noted that such models are very simplified. For example, reaction of a biological axon is chaotic even if the input is periodical. In this aspect the more adequate model of neurons seems to be chaotic. Model of a chaotic neuron can be used as an element of chaotic neural networks. The more adequate results can be obtained if using fuzzy chaotic neural networks, which are closer to biological computation. Fuzzy systems with If-Then rules can model non-linear dynamic systems and capture chaotic attractors easily and accurately. Combination of Fuzzy Logic and Chaos Theory gives us useful tool for building system's chaotic behavior into rule structure. Identification of chaos allows us to determine predicting strategies. If we use a Neural Network Predictor for predicting the system's behavior, the parameters of the strange attractor (in particular fractal dimension) tell us how much data are necessary to train the neural network. The combination of Neurocomputing and Chaotic computing technologies can be very helpful for prediction and control.

The cooperation between these formalisms gives a useful tool for modeling and reasoning under uncertainty in complicated real-world problems. Such cooperation is of particular importance for constructing perception-based intelligent information systems. We hope that the mentioned intelligent combinations will develop further, and the new ones will be proposed. These SC paradigms will form the basis for creation and development of Computational Intelligence.

COGNITIVE MODELS OF GROUND MOTIONS

The existence of numerous databases in the field of civil engineering, and in particular in the field of geotechnical earthquake, has opened new research lines through the introduction of analysis based on soft computing. Three methods are mainly applied in this emerging field: the ones based on the Neural Networks NN, the ones created using Fuzzy Sets FS theory and the ones developed from the Evolutionary Computation [45].

The SC hybrids used in this investigation are directed to tasks of prediction (classification and/or regression). The central objective is obtaining numerical and/or categorical values that mimic input-output conditions from experimentation and in situ measurements and then, through the recorded data and accumulated experience, predict future behaviors. The examples presented herein have been developed by an engineering committee that works for generating useful guidance to geotechnical practitioners with geotechnical seismic design. This effort could help to minimize the perceived significant and undesirable variability within geotechnical earthquake practice. Some urgency in producing the alternative guidelines was seen, after the most recent earthquakes disasters, as being necessary with a desire to avoid a long and protracted process. To this end, a two stage approach was suggested with the first stage being a cognitive interpretation of well-known procedures with appropriate factors for geotechnical design, and a posterior step identifying the relevant philosophy for a new geotechnical seismic design.

Spatial Variation of Soil Dynamic Properties

The spatial variability of subsoil properties constitutes a major challenge in both the design and construction phases of most geo-engineering projects. Subsoil investigation is an imperative step in any civil engineering project. The purpose of an exploratory investigation is to infer accurate information about actual soil and rock conditions at the site. Soil exploration, testing, evaluation, and field observation are well-established and routine procedures that, if carried out conscientiously, will invariably lead to good engineering design and construction. It is impossible to determine the optimum spacing of borings before an investigation begins because the spacing depends not only on type of structure but also on uniformity or regularity of encountered soil deposits. Even the most detail soil maps are not efficient enough for predicting a specific soil property because it changes from place to place, even for the same soil type.

Consequently interpolation techniques have been extensively exploited. The most commonly used methods are kriging and co-kriging but for better estimations they require a great number of measurements available for each soil type, what is generally impossible.

Based on the high cost of collecting soil attribute data at many locations across landscape, new interpolation methods must be tested in order to improve the estimation of soil properties. The integration of GIS and Soft Computing SC offers a potential mechanism to lower the cost of analysis of geotechnical information by reducing the amount of time spent understanding data. Applying GIS to large sites, where historical data can be organized to develop multiple databases for analytical and stratigraphic interpretation, originates the establishment of spatial/chronological efficient methodologies for interpreting properties (soil exploration) and behaviors (in situ measured). GIS-SC modeling/simulation of natural systems represents a new methodology for building predictive models, in this investigation NN and GAs, nonparametric cognitive methods, are used to analyze physical, mechanical and geometrical parameters in a geographical context. This kind of spatial analysis can handle uncertain, vague and incomplete/redundant data when modeling intricate relationships between multiple variables. This means that a NN has not constraints about the spacing (minimum distance) between the drill holes used for building (training) the SC model. The NNs-GAs acts as computerized architectures that can approximate nonlinear functions of several variables, this scheme represent the relations between the spatial patterns of the stratigraphy without restrictive assumptions or excessive geometrical and physical simplifications.

The geotechnical data requirements (geo-referenced properties) for an easy integration of the SC technologies are explained through an application example: a geo-referenced three-dimensional model of the soils underlying Mexico City. The classification/prediction criterion for this very complex urban area is established according to two variables: the cone penetration resistance qc (mechanical property) and the shear wave velocity Vs (dynamic property). The expected result is a 3D-model of the soils underlying the city area that would eventually be improved for a more complex and comprehensive model adding others mechanical, physical or geometrical geo-referenced parameters.

Cone-tip penetration resistances and shear wave velocities have been measured along 16 bore holes spreaded throughout the clay deposits of Mexico City (Figure 2). This information was used as the set of examples inputs (latitude, longitude and depth) → output (qc /Vs). The analysis was carried out in an approximate area of 125 km^2 of Mexico City downtown. It is important to point out that 20% of these patterns (sample points and complete variables information) are not used in the training stage; they will be presented for testing the generalization capabilities of the closed system components (once the training is stopped).

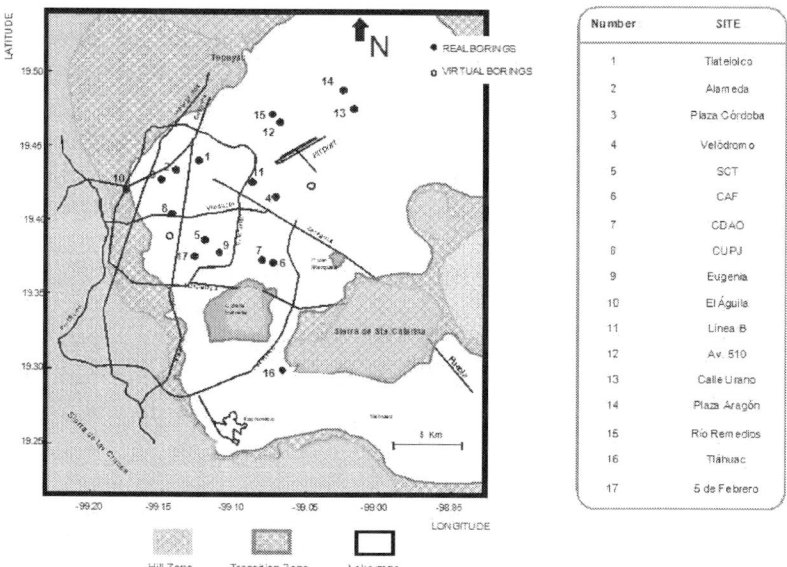

Figure 2: Mexico City Zonation.

In the 3D-neurogenetic analysis, the functions qc = {qc(X,Y,Z)}/ Vs = {Vs(X,Y,Z)} are to be approximated using the procedure outlined below:

1. Generate the database including identification of the site [borings or stations] (X,Y –geographical coordinates, Z –depth, and a CODE –ID number), elevation reference (meters above de sea level, m.a.s.l.), thickness of predetermined structures (layers), and additional information related to geotechnical zoning that could be useful for results interpretation.

2. Use the database to train an initial neural topology whose weights and layers are tuned by an evolutive algorithm (see [46] for details), until the minimum error between calculated and measured values qc = fNN(X,Y,Z)}/Vs = {fNN(X,Y,Z)} is achieved (Figure 3a). The

generalization capabilities of the optimal 3D neural model are tested presenting real work cases (information from borings not included in the training set) to the net. Figure 3b presents the comparison between the measured q_c, V_s values and the NN calculations for testing cases. Through the neurogenetic results for unseen situations we can conclude that the procedure works extremely well in identifying the general trend in materials resistance (stiffness). The correlation between NN calculations and "real" values is over 0.9.

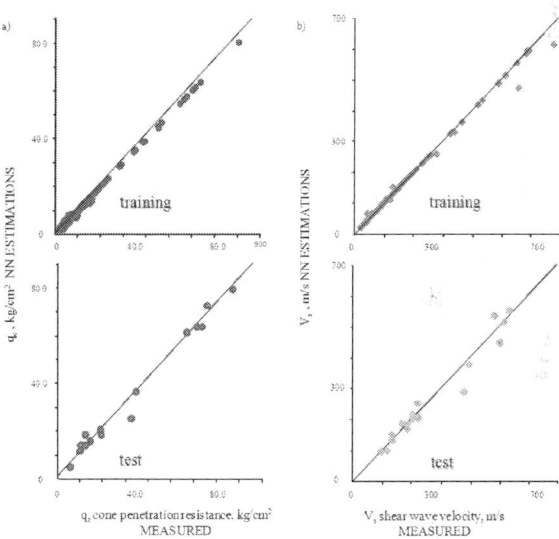

Figure 3: Neural estimations of mechanical and dynamic parameters.

3. For visual environment requirements a grid is constructed using raw information and neurogenetic estimations for defining the spatial variation of properties (Figure 4). The 3D view of the studied zone represents an easier and more understandable engineering system. The 3D neurogenetic-database also permits to display property-contour lines for specific depths. Using the neurogenetic contour maps, the spatial distribution of the mechanical/dynamic variables can be visually appreciated. The 3D model is able to reflect the stratigraphical patterns (Figure 5), indicating that the proposed networks are effective in site characterization with remarkable advantages if comparing with geostatistical approximations: it is easier to use, to understand and to develop graphical user interfaces. The confidence and practical advantages of the defined neurogenetic layers is evident. Precision of

predictions depends on neighborhood structure, grid size, and variance response, but based on the results we can conclude that despite of the grid cell (size) is not too small the spatial correlation extends beyond the training neighborhood, but the higher confidence is obviously only within.

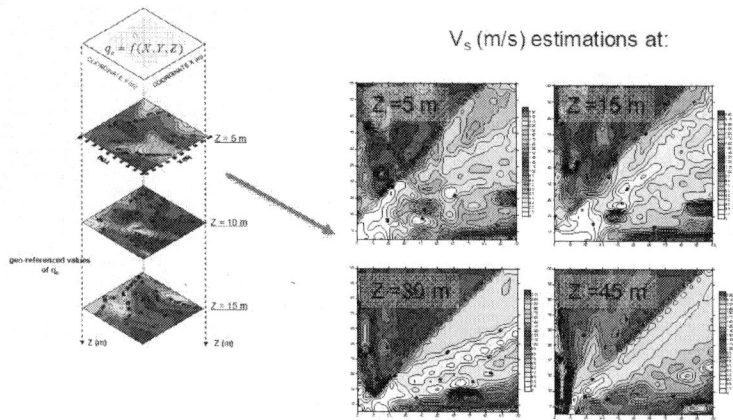

Figure 4: Neural response.

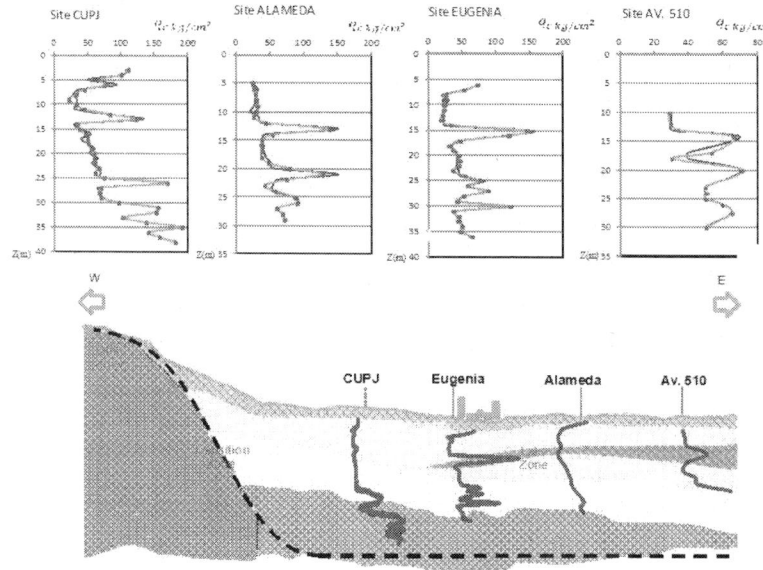

Figure 5: Stratigraphy sequence obtained using the 3D Neural estimations.

Attenuation Laws for Rock Site (Outcropping Motions)

Source, path, and local site response are factors that should be considered in seismic hazard analyses when using attenuation relations. These relations, obtained from statistical regression, are derived from strong motion recordings to define the occurrence of an earthquake with a specific magnitude at a particular distance from the site. Because of the uncertainties inherent in the variables describing the source (e.g. magnitude, epicentral distance, focal depth and fault rupture dimension), the difficulty to define broad categories to classify the site (e.g. rock or soil) and our lack of understanding regarding wave propagation processes and the ray path characteristics from source to site, commonly the predictions from attenuation regression analyses are inaccurate. As an effort to recognize these aspects, multiparametric attenuation relations have been proposed by several researchers [47-53]. However, most of these authors have concluded that the governing parameters are still source, ray path, and site conditions. In this section an empirical NN formulation that uses the minimal information about magnitude, epicentral distance, and focal depth for subduction-zone earthquakes is developed to predict the peak ground acceleration PGA and spectral accelerations Sa at a rock-like site in Mexico City.

The NN model was training from existing information compiled in the Mexican strong motion database. The NN uses earthquake moment magnitudeMw, epicentral distanceED, and focal depth FD from hundreds of events recorded during Mexican subduction earthquakes (Figure 6) from 1964 to 2007. To test the predicting capabilities of the neuronal model, 186 records were excluded from the data set used in the learning phase. Epicentral distance ED is considered to be the length from the point where fault-rupture starts to the recording site, and the focal depth FD is not declared as mechanism classes, the NN should identify the event type through the FD crisp value coupled with the others input parameters [54, 47, 55], The interval of Mw goes from 3 to 8.1 approximately and the events were recorded at near (a few km) and far field stations (about 690 km). The depth of the zone of energy release ranged from very shallow to about 360 km.

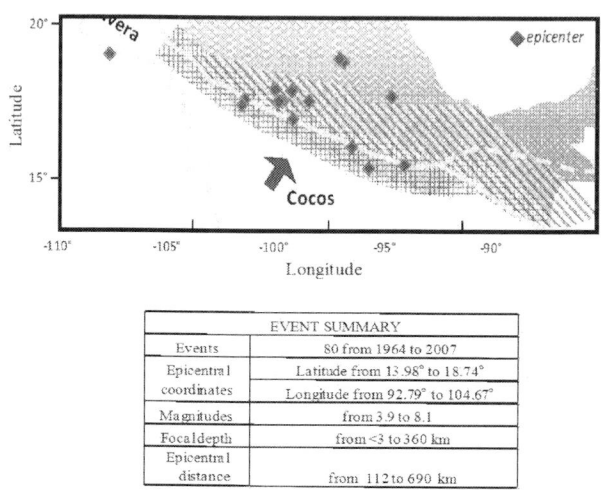

EVENT SUMMARY	
Events	80 from 1964 to 2007
Epicentral coordinates	Latitude from 13.98° to 18.74°
	Longitude from 92.79° to 104.67°
Magnitudes	from 3.9 to 8.1
Focal depth	from <3 to 360 km
Epicentral distance	from 112 to 690 km

Figure 6: Earthquakes characteristics.

Modeling of the data base has been performed using backpropagation learning algorithm. Horizontal (mutually orthogonalPGAh1, N-S component, andPGAh2, E-W component) and vertical components (PGAv) are included as outputs for neural mapping. After trying many topologies, the best horizontal and vertical modules with quite acceptable approximations were the simpler alternatives (BP backpropagation, 2 hidden layers/15 units or nodes each). The neuronal attenuation model for {Mw,ED,FD}→{PGAh1,PGAh2,PGAv} was evaluated by performing testing analyses. The predictive capabilities of the NNs were verified by comparing the estimated PGA's to those induced by the 186 events excluded from the original database (data for training stage). In Figure 7 are compared the computed PGA's during training and testing stages to the measured values. The relative correlation factors (R2≅0.97), obtained in the training phase, indicate that those topologies selected as optimal behave consistently within the full range of intensity, distances and focal depths depicted by the patterns. Once the networks converge to the selected stop criterion, learning is finished and each of these black-boxes become a nonlinear multidimensional functional. Following this procedure 20 NN are trained to evaluate de Sa at different response spectra periods (from T= 0.1 s to T= 5.0 s with DT=0.25 s). Forecasting of the spectral components is reliable enough for practical applications.

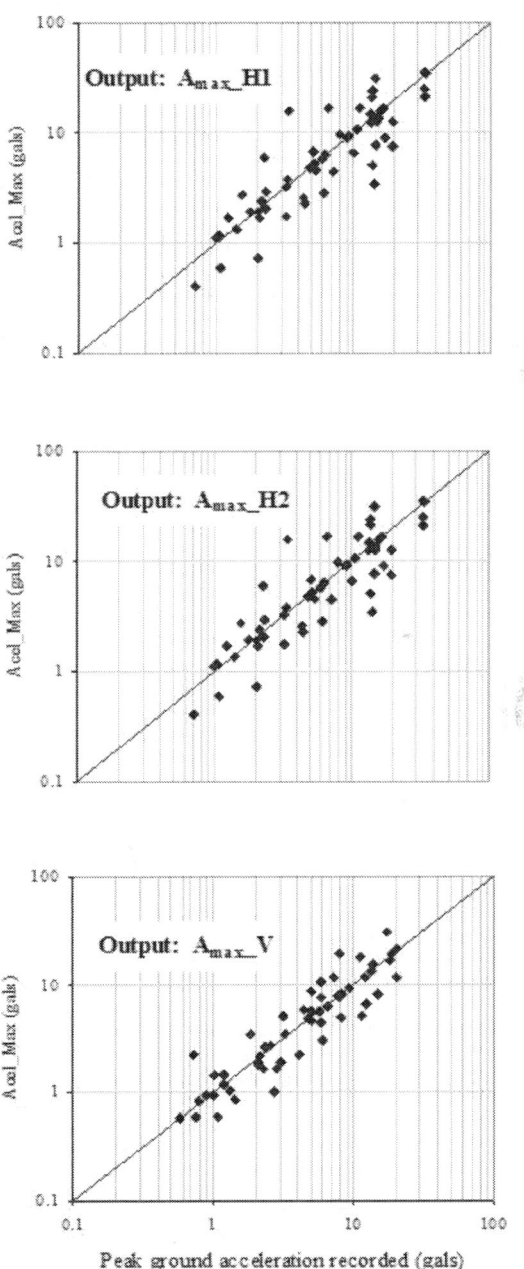

Figure 7: Some examples of measured and NN-estimated PGA values.

In Figure 8 two case histories correspond to large and medium size events are shown, the estimated values obtained for these events using the relationships proposed by Gómez, Ordaz &Tena [56], Youngs et al. [47], Atkinson and Boore [55] –proposed for rock sites– and Crouse et al. [51] – proposed for stiff soil sites– and the predictions obtained with the PGAh1–h2modules are shown. It can be seen that the estimation obtained with Gómez, Ordaz y Tena [56] seems to underestimate the response for the large magnitude event. However, for the lower magnitude event follows closely both the measured responses and NN predictions. Youngs et al. [47] attenuation relationship follows closely the overall trend but tends to fall sharply for long epicentral distances.

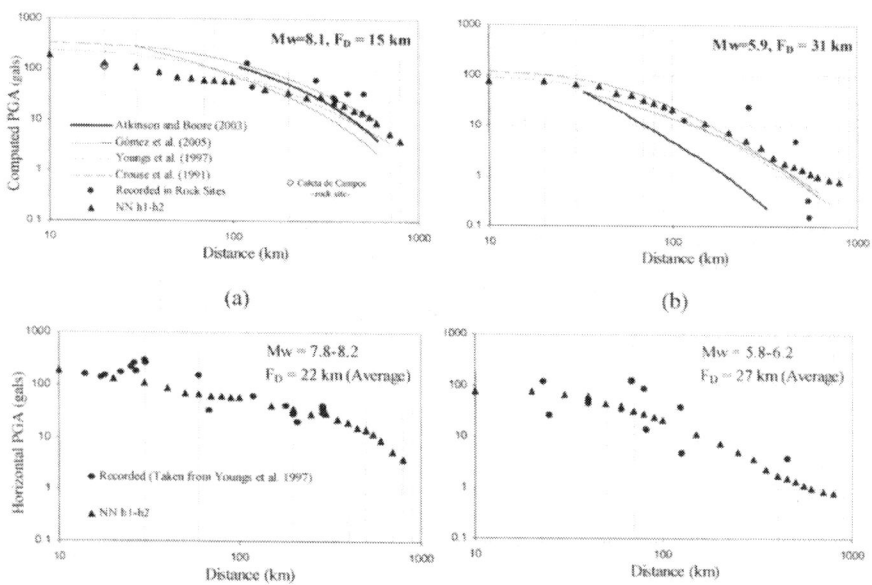

Figure 8: Attenuation laws comparisons.

Furthermore, it should be stressed the fact that, as can be seen in Figure 9 the neural attenuation model is capable to follow the general behavior of the measure data expressed as spectra while the traditional functional approaches are not able to reproduce. A neural sensitivity study for the input variables was conducted for the neuronal modules. The results are strictly valid only for the data base utilized, nevertheless, after several sensitivity analyses conducted changing the database composition, it was found that the following trend prevails; the Mw would be the most relevant parameter then would follow ED coupled with FD. However, for near site events the epicentral distance could become as relevant as the magnitude, particularly, for the vertical component and for minor earthquakes (M low) the FD becomes very transcendental.

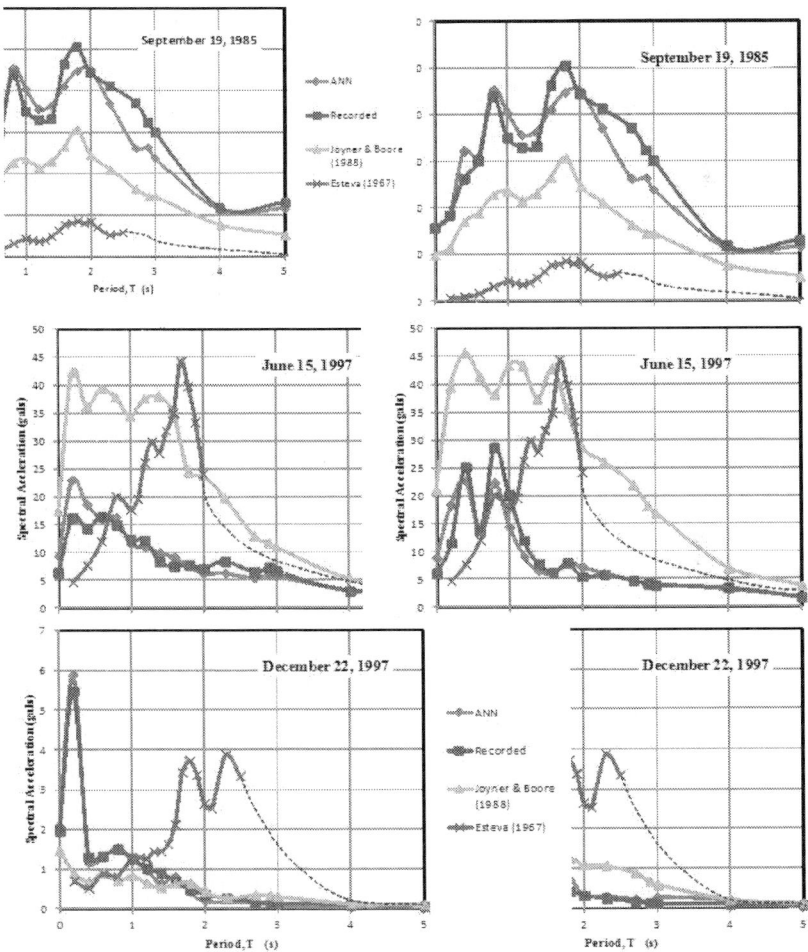

Figure 9: Response Spectra: NN-calculated vs traditional functions.

Through{Mw,ED,FD}→{PGAhi,Sa} mapping, this neuronal approach offers the flexibility to fit arbitrarily complex trends in magnitude and distance dependence and to recognize and select among the tradeoffs that are present in fitting the observed parameters within the range of magnitudes and distances present in data. This approach seems to be a promising alternative to describe earthquake phenomena despite of the limited observations and qualitative knowledge of the recording stations geotechnical site conditions, which leads to a reasoning of a partially defined behavior.

Generation of Artificial Time Series: Accelerograms Application

For nonlinear seismic response analysis, where the superposition techniques do not apply, earthquake acceleration time histories are required as inputs. Virtually all seismic design codes and guidelines require scaling of selected ground motion time histories so that they match or exceed the controlling design spectrum within a period range of interest. Considerable variability in the characteristics of the recorded strong-motions under similar conditions may still require a characterization of future shaking in terms of an ensemble of accelerograms rather than in terms of just one or two "typical" records. This situation has thus created a need for the generation of synthetic (artificial) strong-motion time histories that simulate realistic ground motions from different points of views and/or with different degrees of sophistication. To provide the ground motions for analysis and design, various methods have been developed: i) frequency-domain methods where the frequency content of recorded signals is manipulated [57-60] and ii) time-domain methods where the recorded ground motions amplitude is controlled [61, 62]. Regardless of the method, first, one or more time histories are selected subjectively, and then scaling mechanisms for spectrum matching are applied. This is a trial and error procedure that leads artificial signals very far from real-earthquake time series.

In this investigation a Genetic Generator of Signals is presented. This genetic generator is a tool for finding the coefficients of a pre-specified functional form, which fit a given sampling of values of the dependent variable associated with particular given values of the independent variable(s). When the genetic generator is applied to synthetic accelerograms construction, the proposed tool is capable of i) searching, under specific soil and seismic conditions (within thousands of earthquake records) and recommending a desired subset that better match a target design spectrum, and ii) through processes that mimic mating, natural selection, and mutation, producing new generations of accelerograms until an optimum individual is obtained. The procedure is fast and reliable and results in time series that match any type of target spectrum with minimal tampering and deviation from recorded earthquakes characteristics.

The objective of the genetic generator, when applied to synthetic earthquakes construction, is to produce compatible artificial signals with specific design spectra. In this model specific seismic (fault rupture, magnitude, distance, focal depth) and site characteristics (soil/ rock) are the first set of inputs. They are included to take into consideration that a typical strong motion record consists of a variety of waves whose contribution depends on the earthquake source mechanism (wave path) and its particular characteristics are influenced by the distance between the source and the site, some measure of the size of the earthquake, and the surrounding geology and site conditions; and that the design spectra can be an envelope or integration of many expected ground motions that are possible to occur in certain period of time, or the result of a formulation that involves earthquake magnitude, distance and soil conditions. The second set of inputs consist of the target spectrum, the period range for the matching, lower- and upper-bound acceptable values for scaling signal shape, and a collection of GAs parameters (a population size, number of generations, crossover ratio, and mutation ratio). The output is the more success individual with a chromosome array generated from "real" accelerograms parents (a set of).

The algorithm (see Figure 10) is started with a set of solutions (each solution is called a chromosome). A solution is composed of thousands of components or genes (accelerations recorded at the time), each one encoding a particular trait. The initial solutions (original population) are selected based on the seismic parameters at a site (defined previously by the user): fault mechanism, moment magnitude, epicentral distance, focal depth, geotechnical and geological site classification, depth of sediments. If the user does not have a priori seismic/site knowledge, the genetic generator could select the initial population randomly (Figure 11). Once the model has found the seed-accelerogram(s) or chromosome(s), the space of all feasible solutions can be called accelerograms space (state space). Each point in this search space represents one feasible solution and can be "marked" by its value or fitness for the problem. The looking for a solution is then equal to a looking for some extreme (minimum or maximum) in the space.

According to the individuals' fitness, expressed by difference between the target design spectrum and the chromosome response spectrum, the problem is formulated as the minimization of the error function, Z, between the actual and the target spectrum in a certain period range. Solutions with highest fitness are selected to form new solutions (offspring). During reproduction, the recombination (or crossover) and mutation permits to change the genes (accelerations) from parents (earthquake signals) in some way that the whole new chromosome (synthetic signal) contains the older organisms attributes that assure success. This is repeated until some user's condition (for example number of populations or improvement of the best solution) is satisfied (Figure 12).

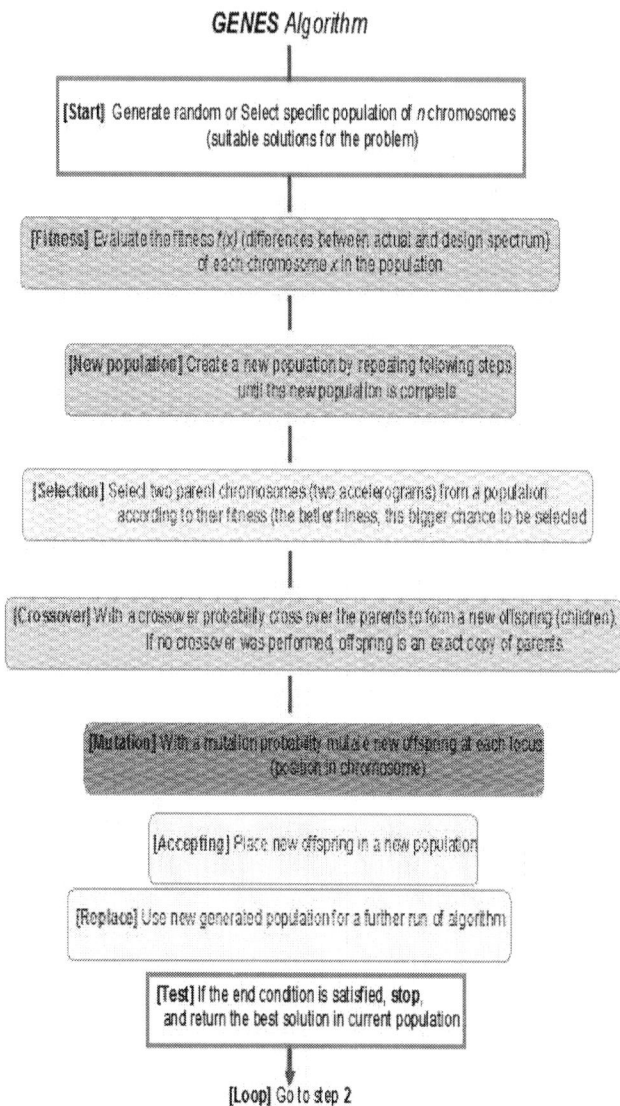

Figure 10: Genetic Generator: flow diagram.

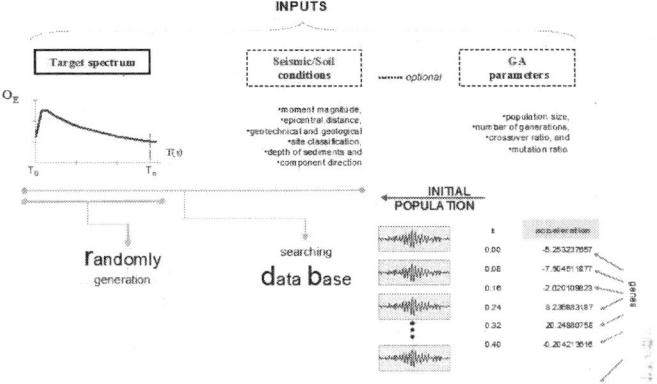

Figure 11: Genetic Generator: working phase diagram.

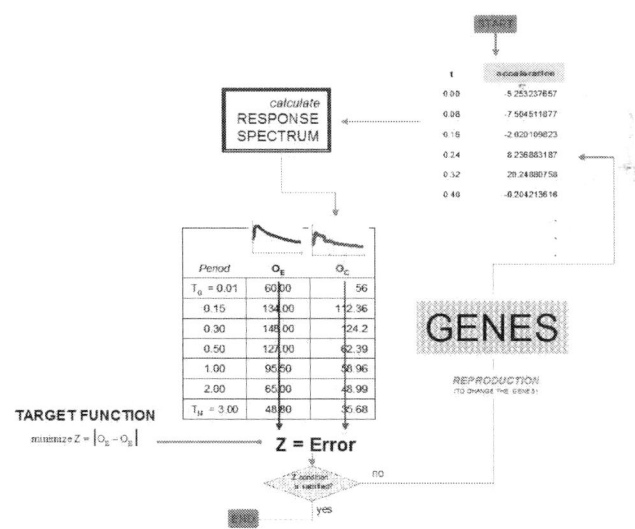

Figure 12: Iteration process of the Genetic Generator.

One of the genetic advantages is the possibility of modifying on line the image of the expected earthquake. While the genetic model is running the user interface shows the chromosome per epoch and its response spectra in the same window, if the duration time, the highest intensities interval or the $\emptyset t$ are not convenient for the user's interests, these values can be modified without retraining or a change on model' structure.

In Figure 13 are shown three examples of signals recovered following this methodology. The examples illustrate the application of the genetic methodology to select any number of records to match a given target spectrum (only the more successful individuals for each target are shown in the figure). It can be noticed the stability of the genetic algorithm in adapting itself to smooth, code or scarped spectrum shapes. The procedure is fast and reliable as results in records match the target spectrum with minimal deviation. The genetic procedure has been applied successfully to generate synthetic ground motions having different amplitudes, duration and combinations of moment magnitude and epicentral distance. Although the variations in the target spectra, the genetic signals maintain the nonlinear and nonstationary characteristics of real earthquakes. It is still under development an additional toolbox that will permit to use advanced signal analysis instruments because, as it has been demonstrated [63] [64], studying nonstationary signals through Fourier or response spectra is not convenient for all applications.

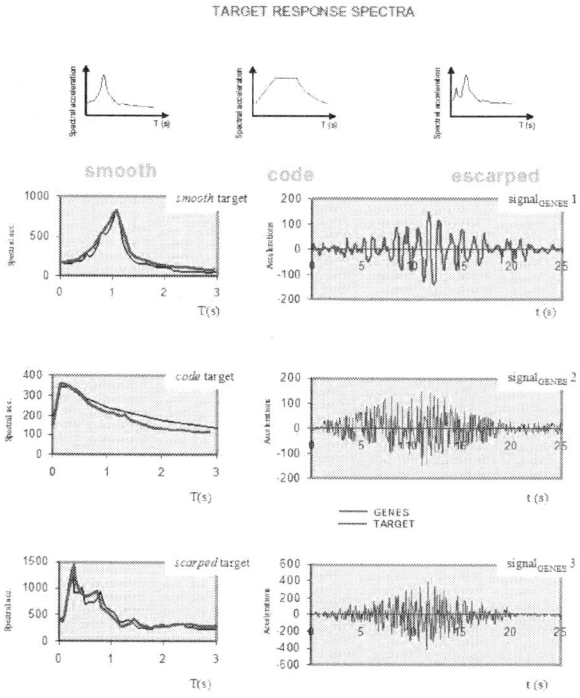

Figure 13: Some Generator results: accelerograms application.

Effects of Local Site Conditions on Ground Motions

Geotechnical and structural engineers must take into account two fundamental characteristics of earthquake shaking: 1) how ground shaking propagates through the Earth, especially near the surface (site effects), and 2) how buildings respond to this ground motion. Because neither characteristic is completely understood, the seismic phenomenon is still a challenging research area.

Site effects play a very important role in forecasting seismic ground responses because they may strongly amplify (or deamplify) seismic motions at the last moment just before reaching the surface of the ground or the basement of man-made structures. For much of the history of seismological research, site effects have received much less attention than they should, with the exception of Japan, where they have been well recognized through pioneering work by Sezawa and Ishimoto as early as the 1930's [65]. The situation was drastically changed by the catastrophic disaster in Mexico City during the Michoacan, Mexico earthquake of 1985, in which strong amplification due to extremely soft clay layers caused many high-rise buildings to collapse despite their long distance from the source. The cause of the astounding intensity and long duration of shaking during this earthquake is not well resolved yet even though considerable research has been conducted since then, however, there is no room for doubt that the primary cause of the large amplitude of strong motions in the soft soil (lakebed) zone relative to those in the hill zone is a site effect of these soft layers.

The traditional data-analysis methods to study site effects are all based on linear and stationary assumptions. Unfortunately, in most soil systems, natural or manmade ones, the data are most likely to be both nonlinear and nonstationary. Discrepancies between calculated responses (using code site amplification factors) and recent strong motion evidence point out serious inaccuracies may be committed when analyzing amplification phenomena. The problem might be due partly because of the lack of understanding regarding the fundamental causes in soil response but also a consequence of the distorted soil amplification quantification and the incomplete characterization of nonlinearity-induced nonstationary features exposed in motion recordings [66]. The objective of this investigation is to illustrate a manner in which site effects can be dealt with for the case of Mexico City soils, making use of response spectra calculated from the motions recorded at different sites during extreme and minor events (see Figure 6). The variations in the spectral shapes, related to local site conditions, are used to feed a multilayer neural network that represent a very advantageous nonlinear-amplification relation. The database is composed by registered information earthquakes affecting Mexico City originated by different source mechanisms.

The most damaging shocks, however, are associated to the subduction of the Cocos Plate into the Continental Plate, off the Mexican Pacific Coast. Even though epicentral distances are rather large, these earthquakes have recurrently damaged structures and produced severe losses in Mexico City. The singular geotechnical environment that prevails in Mexico City is the one most important factor to be accounted for in explaining the huge amplification of seismic movements [67-70]. The soils in Mexico City were formed by the deposition into a lacustrine basin of air and water transported materials. From the view point of geotechnical engineering, the relevant strata extend down to depths of 50 m to 80 m, approximately. The superficial layers formed the bed of a lake system that has been subjected to dessication for the last 350 years. Three types of soils may be broadly distinguished: in Zone I, firm soils and rock-like materials prevail; in Zone III, very soft clay formations with large amounts of microorganisms interbedded by thin seams of silty sands, fly ash and volcanic glass are found; and in Zone II, which is a transition between zones I and III, sequences of clay layers an coarse material strata are present (Figure 14).

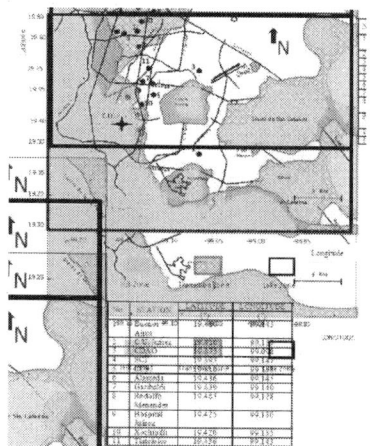

Figure 14: Accelerographic stations used in this study.

Due to space limitations, reference is made only to two seismic events: the June 15, 1999 and the October 9, 1995. This module was developed based in a previous study (see section 4.2 of this Chapter) where the effect of the parametersED, FD and Mw on the ground motion attenuation from epicentre to the site, were found to be the most significant [71]. The recent disaster experience showed that the imprecision that is inherent to most variables measurements or estimations makes crucial the consideration of subjectivity to evaluate and to derive numerical conclusions according to the phenomena behavior. The neuronal training process starts with the training of four input variables booked:ED, FD and Mw. The output linguistic variables are PGAh1 (peak ground acceleration horizontal, component 1) and ,PGAh2 (peak ground acceleration, horizontal component 2) registered in a rock-like site in Zone I .The second training process is linked *feed-forward* with the previous module (PGA for rock-like site) and the new seismic inputs are Seismogenic Zone and PGA_{rock} and the Latitude and Longitude coordinates are the geo-referenced position needed to draw the deposition variation into the basin. This neuro-training runs one step after the first training phase and until the minimum difference between theSa and the neuronal calculations is attained. In Figure 15 some results from training and testing modes are shown.

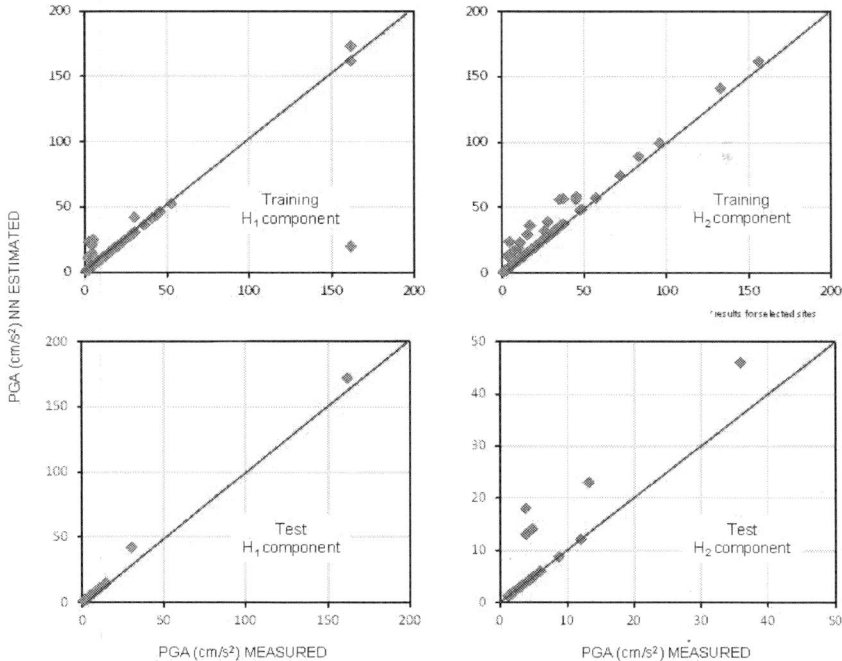

Figure 15: Neural estimations for PGA in Lake Zone sites.

This second NN represents the geo-referenced amplification ratio that take into consideration the topographical, geotechnical and geographical conditions, implicit in the recorded accelerograms.The results of these two

NNs are summarized in Figure 16. These graphs show the predicting capabilities of the neural system comparing the measured values with those obtained in neural-working phase. It can be observed a good correspondence throughout the full distance and magnitude range for the seismogenic zones considered in this study for the whole studied area (Lake Zone).

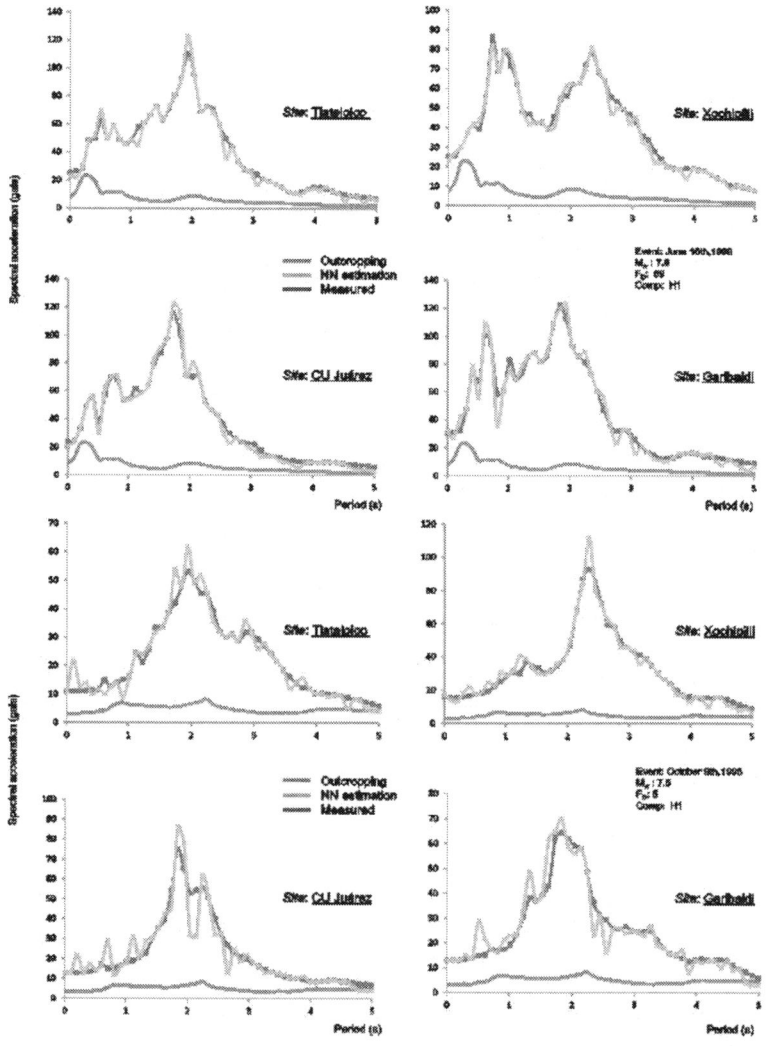

Figure 16: Spectral accelerations in some Lake-Zone sites: measured vs NN.

Liquefaction Phenomena: Potential Assessment and Lateral Displacements Estimation

Over the past forty years, scientists have conducted extensive research and have proposed many methods to predict the occurrence of liquefaction. In the beginning, undrained cyclic loading laboratory tests had been used to evaluate the liquefaction potential of a soil [72] but due to difficulties in obtaining undisturbed samples of loose sandy soils, many researchers have preferred to use *in situ* tests [73]. In a semi-empirical approach the theoretical considerations and experimental findings provides the ability to make sense out of the field observations, tying them together, and thereby having more confidence in the validity of the approach as it is used to interpolate or extrapolate to areas with insufficient field data to constrain a purely empirical solution. Empirical field-based procedures for determining liquefaction potential have two critical constituents: i) the analytical framework to organize past experiences, and ii) an appropriate *in situ* index to represent soil liquefaction characteristics. The original simplified procedure [74] for estimating earthquake-induced cyclic shear stresses continues to be an essential component of the analysis framework. The refinements to the various elements of this context include improvements in the in-situ index tests (e.g., SPT, CPT, BPT,Vs), and the compilation of liquefaction/no-liquefaction cases.

The objective of the present study is to produce an empirical machine learning ML method for evaluating liquefaction potential. ML is a scientific discipline concerned with the design and development of algorithms that allow computers to evolve behaviours based on empirical data, such as from sensor data or databases. Data can be seen as examples that illustrate relations between observed variables. A major focus of ML research is to automatically learn to recognize complex patterns and make intelligent decisions based on data; the difficulty lies in the fact that the set of all possible behaviours given all possible inputs is too large to be covered by the set of observed examples (training data). Hence the learner must generalize from the given examples, so as to be able to produce a useful output in new cases. In the following two ML tools, Neural Networks NN and Classification Trees CTs, are used to evaluate liquefaction potential and to find out the liquefaction control parameters, including earthquake and soil conditions. For each of these parameters, the emphasis has been on developing relations that capture the essential physics while being as simplified as possible. The proposed cognitive environment permits an improved definition of i) seismic loading or cyclic stress ratio CSR, and ii) the *resistance* of the soil to triggering of liquefaction or cyclic resistance ratio CRR.

The factor of safety FS against the initiation of liquefaction of a soil under a given seismic loading is commonly described as the ratio of cyclic resistance ratio (CRR), which is a measure of liquefaction resistance, over cyclic stress ratio (CSR), which is a representation of seismic loading that causes liquefaction, symbolically, FS=CRR/CSR. The reader is referred to Seed and Idriss [74], Youd et al.[75], and Idriss and Boulanger [76] for a historical perspective of this approach. The term CSR CSR=f(0.65, ɣvo,amax, ɣ'vo,rd,MSF)is function of the vertical total stress of the soil ɣvo at the depth considered, the vertical effective stress ɣ'vo, the peak horizontal ground surface acceleration amax, a depth-dependent shear stress reduction factor rd (dimensionless), a magnitude scaling factor MSF (dimensionless). For CRR, different in situ-resistance measurements and overburden correction factors are included in its determination; both terms operate depending of the geotechnical conditions. Details about the theory behind this topic in Idriss and Boulanger, [76] and Youd et al. [75].

Many correction/adjustment factors have been included in the conventional analytical frameworks to organize and to interpret the historical data. The correction factors improve the consistency between the geotechnical/seismological parameters and the observed liquefaction behavior, but they are a consequence of a constrained analysis space: a 2D plot [CSR *vs.* CRR] where regression formulas (simple equations) intend to relate complicated nonlinear/multidimensional information. In this investigation the ML methods are applied to discover unknown, valid patterns and relationships between geotechnical, seismological and engineering descriptions using the relevant available information of liquefaction phenomena (expressed as empirical prior knowledge and/or input-output data). These ML techniques "work" and "produce" accurate predictions based on few logical conditions and they are not restricted for the mathematical/analytical environment. The ML techniques establish a *natural* connection between experimental and theoretical findings.

Following the format of the simplified method pioneered by Seed and Idriss [74], in this investigation a nonlinear and adaptative *limit state* (a fuzzy-boundary that separates liquefied cases from nonliquefied cases) is proposed (Figure 17). The database used in the present study was constructed using the information included in Table 2 and it was compiled by Agdha et al., [77], Juang et al., [78], Juang [79], Baziar, [80] and Chern and Lee [81]. The cases are derived from cone penetration tests CPT, and shear wave velocities Vs measurements and different world seismic conditions (U.S., China, Taiwan, Romania, Canada and Japan). The soils types ranges from clean sand and silty sand to silt mixtures (sandy and clayey silt). Diverse geological and geomorphological characteristics are included. The reader is referred to the citations in Table 2 for details.

i)

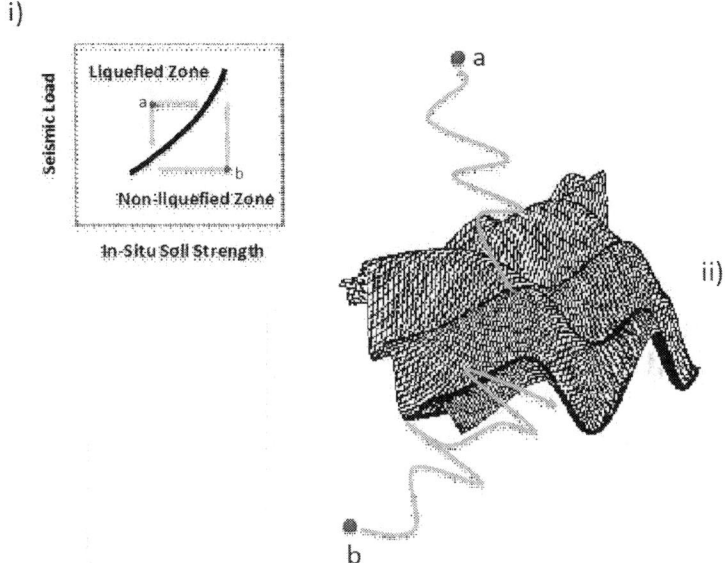

ii)

Figure 17: An schematic view of the nonlinear-liquefaction boundary.

The ML formulation uses Geotechnical (qc, Vs, Unit weight, Soil Type, Total vertical stresses, Effective vertical stresses, Geometrical (Layer thickness, Water Level Depth, Top Layer Depth) andSeismological (Magnitude, PGA) input parameters and the output variable is "Liquefaction?" and it can take the values "YES/NO" (Figure 17). Once the NN is trained the number of cases that was correctly evaluated was 100% and applied to "unseen" cases (separated for testing) less than 10% of these examples were not fitted. The CT has a minor efficiency during the training showing 85% of cases correctly predicted, but when the CT runs on the unseen patterns its capability is not diminished and it asserts the same proportion. From these findings it is concluded that the neuro system is capable of predicting the in situ measurements with a high degree of accuracy but if improvement of knowledge is necessary or there are missed, vague even contradictory values in the analyzed case, the CT is a better option.

Table 2: Database for liquefaction analysis.

Set	Input Parameters	Number .of Patterns	Ref.
A	Z, Z_{NAF}; H, Soil Class, Geomorphological units, Geological units, Site amplification, a_{max}	56	Fatemi-Agdha et al., 1988
B	Z, q_c, F_s, σ_0, σ_0', a_{max}, M	21	Juang et al., 1999
C	Z, q_c, F_s, σ_0, σ_0', a_{max}, M	242	Juang, 2003
D	D_{50}, a_{max}, σ_0', σ_0, M, F_s, q_c, SPT, Z	170	Baziar, 2003
E	M, σ_0, σ_0', q_c, a_{max}	466	Chern and Lee, 2009
F	Z_{NAF}, Z, H, σ_0, σ_0', Soil Class, V_s	80	Andrus and Stokoe, 1997; 2000
	Total:	**1035**	

Figure 18 shows the pruned liquefaction trees (two, one runs using qc values and the other through theVs measurements) with YES/NO as terminal nodes. In the Figure 19, some examples of tree reading are presented. The trees incorporate soil type dependence through the resistance values (qc, andVs) and fine content, and it is not necessary to label the material as "sand" or "silt". The most general geometrical branches that split the behaviors are the Water table depth and the Layer thickness but only when the soil description is based on Vs, when qc, serves as rigidity parameter this geometrical inputs are not explicit exploited. This finding can be related to the nature of the measurement: the cone penetration value contains the effect of the saturated material while the shear wave velocities need the inclusion of this situation explicitly. Without potentially confusing regression strategies, the liquefaction trees results can be seen as an indication of how effectively the ML model maps the assigned predictor variables to the response parameter. Using data from all regions and wide parameters ranges, the prediction capabilities of the neural network and classification trees are superior to many other approximations used in common practice, but the most important remark is the generation of meaningful clues about the reliability of physical parameters, measurement and calculation process and practice recommendations.

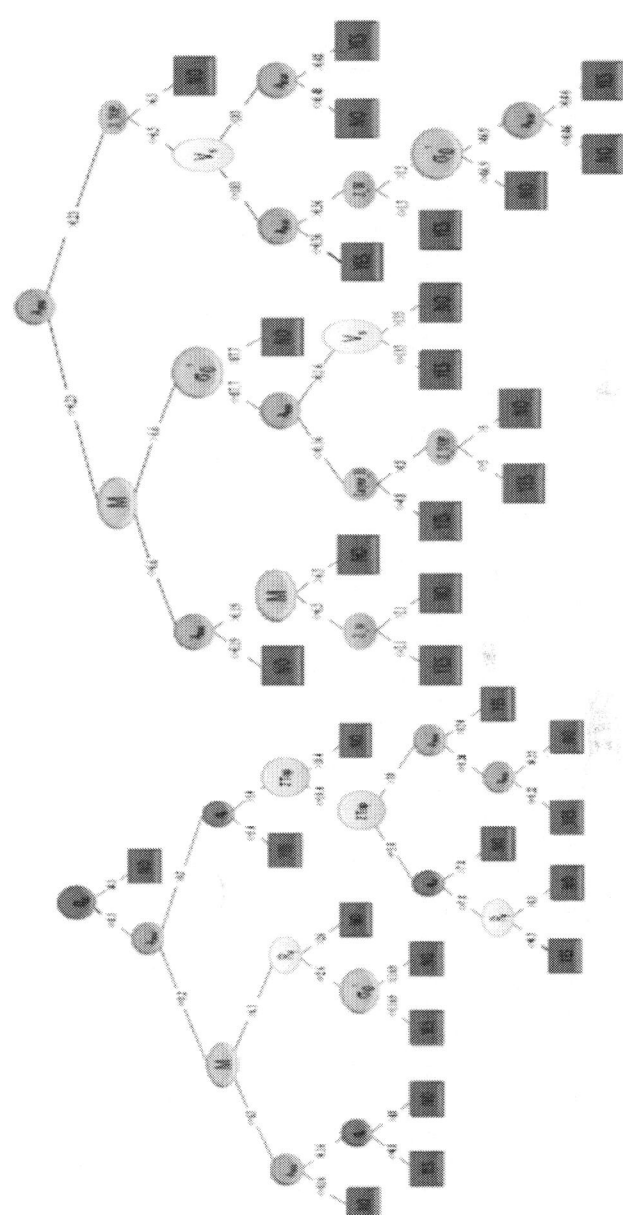

Figure 18: Classification tree for liquefaction potential assessment.

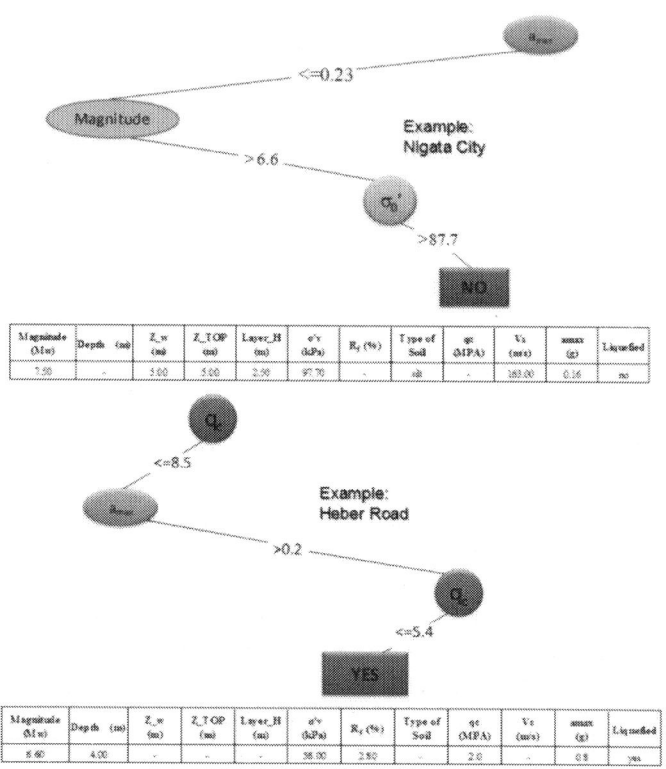

Figure 19: CT classification examples.

The intricacy and nonlinearity of the phenomena, an inconsistent and contradictory database, and many subjective interpretations about the observed behavior, make SC an attractive alternative for estimation of liquefaction induced lateral spread. NEFLAS [82], NEuroFuzzy estimation of liquefaction induced LAteral Spread, profits from fuzzy and neural paradigms through an architecture that uses a fuzzy system to represent knowledge in an interpretable manner and proceeds from the learning ability of a neural network to optimize its parameters. This blending can constitute an interpretable model that is capable of learning the problem-specific prior knowledge.

NEFLAS is based on the Takagi-Sugeno model structure and it was constructed according the information compiled by Bartlett and Youd [83] and extended later by Youd et al. [84]. The output considered in NEFLAS is the horizontal displacements due to liquefaction, dependent of moment magnitude, the PGA, the nearest distance from the source in kilometers; the free face ratio, the gradient of the surface topography or the slope of the liquefied layer base, the cumulative thickness of saturated cohesionless sediments with number of blows (modified by overburden and energy delivered to the standard penetration probe, in this case 60%) , the average of fines content, and the mean grain size.

One of the most important NEFLAS advantages is its capability of dealing with the imprecision, inherent in geoseismic engineering, to evaluate concepts and derive conclusions. It is well known that engineers use words to classify qualities ("strong earthquake", "poor graduated soil" or "soft clay" for example), to predict and to validate "first principle" theories, to enumerate phenomena, to suggest new hypothesis and to point the limits of knowledge. NEFLAS mimics this method. See the technical quantity "magnitude" (earthquake input) depicted in Figure 20. The degree to which a crisp magnitude belongs to LOW, MEDIUM or HIGH linguistic label is called the degree of membership. Based on the figure, the expression, "the magnitude is LOW" would be true to the degree of 0.5 for aMwof 5.7. Here, the degree of membership in a set becomes the degree of truth of a statement.

On the other hand, the human logic in engineering solutions generates sets of behavior rules defined for particular cases (parametric conditions) and supported on numerical analysis. In the neurofuzzy methods the human concepts are re-defined through a flexible computational process (training) putting (empirical or analytical) knowledge into simple "if-then" relations (Figure 20). The fuzzy system uses 1) variables composing the antecedents (premises) of implications; 2) membership functions of the fuzzy sets in the premises, and 3) parameters in consequences for finding simpler solutions with less design time.

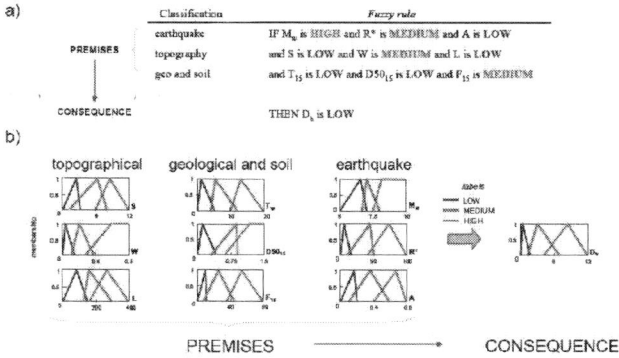

Figure 20: Neurofuzzy estimation of lateral spread.

NEFLAS considers the character of the earthquake, topographical, regional and geological components that influence lateral spreading and works through three modules: Reg-NEFLAS, appropriate for predicting horizontal displacements in geographic regions where seismic hazard surveys have been identified; Site- NEFLAS, proper for predictions of horizontal displacements for site-specific studies with minimal data on geotechnical conditions and Geotech-NEFLAS allows more refined predictions of horizontal displacements when additional data is available from geotechnical soil borings. The NEFLAS execution on cases not included in the database (Figure 21.b and Figure 21.c) and its higher values of correlation when they are compared with evaluations obtained from empirical procedures permit to assert that NEFLAS is a powerful tool, capable of predicting lateral spreads with high degree of confidence.

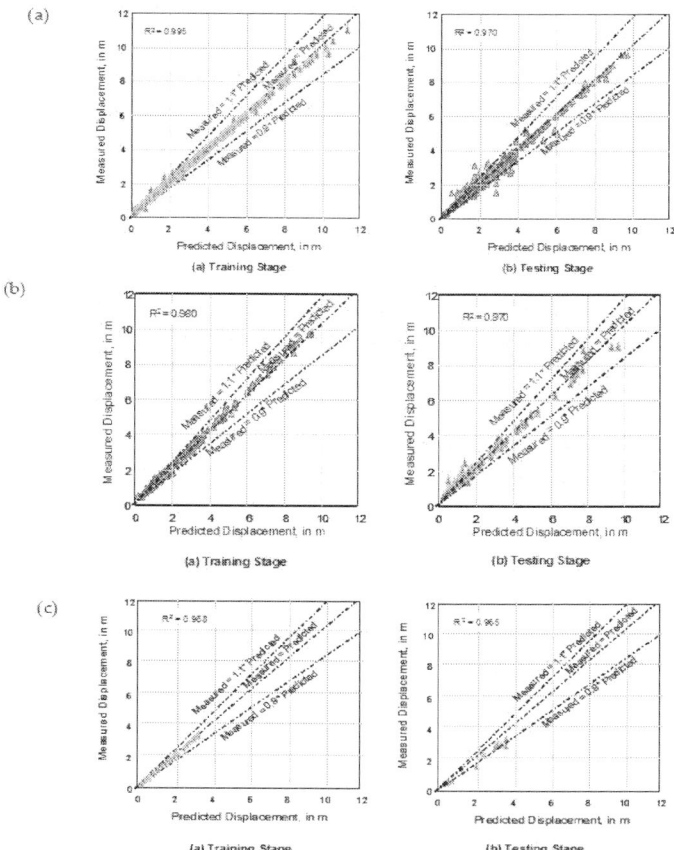

Figure 21: NN estimations vs measured displacements for a) the whole data set, b)Niigata Japan, c) San Francisco USA cases.

CONCLUSIONS

Based on the results of the studies discussed in this paper, it is evident that cognitive techniques perform better than, or as well as, the conventional methods used for modeling complex and not well understood geotechnical earthquake problems. Cognitive tools are having an impact on many geotechnical and seismological operations, from predictive modeling to diagnosis and control.

The hybrid *soft* systems leverage the tolerance for imprecision, uncertainty, and incompleteness, which is intrinsic to the problems to be solved, and generate tractable, low-cost, robust solutions to such problems. The synergy derived from these hybrid systems stems from the relative ease with which we can translate problem domain knowledge into initial model structures whose parameters are further tuned by local or global search methods. This is a form of methods that do not try to solve the same problem in parallel but they do it in a mutually complementary fashion. The push for low-cost solutions combined with the need for intelligent tools will result in the deployment of hybrid systems that efficiently integrate reasoning and search techniques.

Traditional earthquake geotechnical modeling, as physically-based (or knowledge-driven) can be improved using soft technologies because the underlying systems will be explained also based on data (CC data-driven models). Through the applications depicted here it is sustained that cognitive tools are able to make abstractions and generalizations of the process and can play a complementary role to physically-based models.

REFERENCE

1. W.D.L.Finn State of the art of geotechnical earthquake engineering practice Soil Dynamics and Earthquake Engineering 20 (2000Elsevier

2. N. A. R. R. Abrahamson, A. Youngs, algorithm. stable, regression. for, using. analysis, random. the, model. effects, Bull Seismol Soc Am 1992

3. P. G. Somerville, T. Sato, Correlation of rise time with the style-offaulting factor in strong ground motions. Seismol Res Lett 1998pp (abstract).

4. Somerville PG, Greaves RL.Strong ground motions of the Kobe, Japan earthquake of January 17, 1995and development of a model of forward rupture directivity applicable in California. Proc. Western Regional Tech. Seminar on Earthquake Eng. for Dams. Assoc. of State Dam Safety Oficials, Sacramento, CA, April 11±12, 1996.

5. Abrahamson NA, Silva WJ.Empirical duration relations for shallow crustal earthquakes. Written communication, 1997

6. R. W. Graves, A. Pitarka, P. G. Somerville, Ground motion amplication in the Santa Monica area: effects of shallow basin edge structure. Submitted for publication.

7. P. G. Somerville, art. Emerging, ground. earthquake, In. motion, P. Dakoulas, M. Yegian, R. D. Holtz, editors, Proc Geotechnical Earthquake Engineering in Soil Dynamics III, Geotechnical Special Publication 751ASCE, 19981

8. Abrahamson NA.Spatial variation of multiple support inputs. Proc. First US Symp. Seism. Eval. Retrofit Steel Bridges, UC, Berkeley, October 18, 1993

9. F. Naeim, M. Lew, On the use of design spectrum compatible motions. Earthquake Spec 1995II(1):111±28.

10. Boulanger R.W. and Idriss, I.M.2006Liquefaction Susceptibility Criteria for Silts and Clays," Journal of Geotechnical and Geoenvironmental Engineering, 132 (11), 1413 EOF1426 EOFpp.

11. Boulanger R.W. and Idriss, I.M.2007Evaluation of Cyclic Softening in Silts and Clays", Journal of Geotechnical and Geoenvironmental Engineering, 133 (6), 641 EOF652 EOFpp.

12. Youd and Idriss, NCEER. Proceedings, Workshop on Evaluation of Liquefaction Resistance of Soils.Technical Report NCCER-97970022National Center for Earthquake Engineering Research, University of Buffalo, Buffalo, New York, 1997

13. H. B. Seed, K. Tokimatsu, L. F. Harder, R. M. Chung, of. S. P. T. Influence, in. procedures, liquefaction. soil, evaluations. resistance, J Geotech Engng 19851425 EOF

14. Robertson PK, Fear CE.Liquefaction of sands and its evaluation. Proceedings, 1st Int. Conf. on Earthquake Geotechnical Engineering, Tokyo, Japan, 1995

15. Ambraseys NN. Engineering seismology.Earthquake Engng Struct Dynam 1988

16. Y. Wang, 2008aOn Contemporary Denotational Mathematics for Computational Intelligence. In Transactions of Computational Science (2629New York: Springer.

17. Y. Wang, 2009aOn Abstract Intelligence: Toward a Unified Theory of Natural, Artificial, Machinable, and Computational Intelligence. International Journal of Software Science and Computational Intelligence, 1 EOF17 EOF

18. Y. Wang, 2009bOn Cognitive Computing. International Journal of Software Science and Computational Intelligence, 1(3), 1-15.

19. A. M. Turing, 1950Computing Machinery and Intelligence. Mind, 59433460

20. Neumann. J. Von, 1946The Principles of Large-Scale Computing Machines. Reprinted in Annals of History of Computers, 3(3), 263-273.

21. Neumann. J. Von, 1958The Computer and the Brain. New Haven: Yale Univ. Press.

22. J. L. Gersting, 1982Mathematical Structures for Computer Science. San Francisco: W. H. Freeman & Co.

23. D. Mandrioli, C. Ghezzi, 1987Theoretical Foundations of Computer Science. New York: John Wiley & Sons.

24. H. R. Lewis, C. H. Papadimitriou, 1998Elements of the Theory of Computation, 2nd ed. Englewood Cliffs, NJ: Prentice Hall International.

25. J. Kephart, D. Chess, 2003The Vision of Autonomic Computing. IEEE Computer, 26(1), 41 EOF50 EOF

26. IBM2006Autonomic Computing White Paper. An Architectural Blueprint for Autonomic Computing, 4th ed., June, (137

27. Y. Wang, 2004Keynote: On Autonomic Computing and Cognitive Processes. Proc. 3rd IEEE International Conference on Cognitive Informatics (ICCI'04), Victoria, Canada, IEEE CS Press, (34

28. Y. Wang, 2007aJuly). Software Engineering Foundations: A Software Science Perspective. CRC Book Series in Software Engineering, Vol. II, Auerbach Publications, NY.

29. B. Bouchon-Meunier, R. Yager, L. Zadeh, 1995Fuzzy Logic and SoftComputing. World Scientific, Singapore.

30. J. C. Bezdek, is. a. What, intelligence?. computational, Zurada. J. M. In, I. I. R. J. Marks, C. J. Robinson, eds.) Computational Intelligence: Imitating Life, 112IEEE Press, Los Alamitos (1994

31. L. A. Zadeh, Sets. Fuzzy, Information and Control 8, 338-353 (1965

32. Zadeh L.A.The roles of fuzzy logic and soft computing in the conception, design and deployment of intelligent systems. BT Technol J. 14199443236

33. R. A. Aliev, R. R. Aliev, Computing. Soft, I. I. I. I. I. I. volumes, A. S. O. A. Baku, Press, 1997-1998in Russian).

34. R. Aliev, K. Bonfig, F. Aliew, Computing. Soft, Verlag. Berlin, Technic, 2000

35. Zadeh L.A. Foreword. In Proc.First European Congress on Intelligent Techniques and Soft Computing- EUFIT'95, page VII, 1995

36. L. A. Zadeh, Soft Computing and Fuzzy Logic. IEEE Software 11199464858

37. R. A. Aliev, Expert. Fuzzy, In. Systems, F. Aminzadeh, M. Jamshidi, S. O. F. T. C. O. M. P. U. T. I. N. G. (eds, Logic. Fuzzy, Networks. Neural, Artificial. Distributed, 9910 . Intelligence.pages, N. , NJ: PTR Prentice Hall, 1994

38. Zadeh L.A. Fuzzy logic, Neural Networks and Soft Computing .Comm of ACM 37199437784

39. Welstead S.T. (ed) Neural Networks and Fuzzy Logic Applications in C/C++, Professional Computing.NY: John Wiley, 1994

40. Yager R.R. and Zadeh L.A. (eds) Fuzzy sets, neural networks and Soft Computing.NY: VAN Nostrand Reinhold , 1994

41. Nauck D., Klawonn F., and Kruse R., Foundations of Neuro-Fuzzy Systems.NY: John Wiley and Sons, 1997.

42. Mohamad H.Hassoun, Fundamentals of artificial neural networks.Cambridge: MIT Press, 1995

43. S. Haykin, Networks. A. Neural, Foundation. Comprehensive, Marmillau, I. E. E. E. Computer, Society, 1994

44. Goldberg D.E., Genetic algorithms in search, optimization and machine learning.Reading, MA: Addison-Wesley, 1989

45. Arciszewski, and De Jong K.A. (2001Evolutionary computation in civil engineering: research frontiers. Eight International Conference on Civil and Structural Engineering Computing, (Topping B. H. V., ed.), Saxe-Coburg Publications, Eisenstadt, Vienna, Austria.

46. K. Miettinen, P.and. Neittaanmaki, J. Periaux, 1999Evolutionary Algorithms in Engineering and Computer Science : Recent Advances in Genetic Algorithms, Evolution Strategies, Evolutionary Programming, John Wiley & Sons Ltd., pps. 483. 0-47199-902-4

47. R. R. Youngs, S. J. Chiou, W. J. Silva, J. R. Humphrey, 1997Strong ground motion attenuation relationships for subduction zone earthquakes. Seismol. Res. Lett., (68) 1, 58 EOF73 EOF

48. J. G. Anderson, 1997Nonparametric description of peak acceleration above a subduction thrust. Seismol. Res. Lett., (68) 1, 86 EOF93 EOF

49. C. B. Crouse, 1991Ground motion attenuation equations for earthquakes on the Cascadia subduction zone. Earth. Spectra, 7210236

50. S. K. Singh, M. Ordaz, M. Rodríguez, R. Quaas, V. Mena, M. Ottaviani, J. G. Anderson, D. Almora, 1989Analysis of near-source strong motion recordings along the Mexican subduction zone. Bull. Seism. Soc. Am., 7916971717

51. C. B. Crouse, Y. K. Vyas, B. A. Schell, 1988Ground motions from subduction-zone earthquakes. Bull. Seism. Soc. Am.,78125

52. S. K. Singh, E. Mena, R. Castro, C. Carmona, 1987Empirical prediction of ground motion in Mexico City from coastal earthquakes. Bull. Seism. Soc. Am., 7718621867

53. K. Sadigh, 1979Ground motion characteristics for earthquakes originating in subduction zones and in the western United States. Proc. Sixth Pan Amer. Conf., Lima, Peru.

54. B. F. Tichelaar, L. J. Ruff, 1993Depth of seismic coupling along subduction zones. J. Geophys. Res., 9820172037

55. G. M. Atkinson, D. M. Boore, 2003Empirical ground-motion Relations for Subduction-Zone Earthquakes and Their Applications to Cascadia and other regions. Bull. Seism. Soc. Am., 93417031729

56. S. C. Gómez, M. Ordaz, C. Tena, 2005Leyes de atenuación en desplazamiento y aceleración para el diseño sísmico de estructuras con aislamiento en la costa del Pacífico. Memorias del XV Congreso Nacional de Ingeniería Sísimica, México, Nov. A-II-02

57. D. Gasparini, E. H. Vanmarcke, 1976SIMQKE: A Program for Artificial Motion Generation, Department of Civil Engineering, Massachusetts Institute of Technology, Cambridge, MA.

58. W. J. Silva, K. Lee, 1987WES RASCAL code for synthesizing earthquake ground motions." State-of-the-Art for Assessing Earthquake Hazards in the United States, Report 24, U.S. Army Engineers Waterways Experiment Station, Misc. Paper S-731

59. B. A. Bolt, N. J. Gregor, 1993Synthesized Strong Ground Motions for the Seismic Condition Assessment of the Eastern Portion of the San Francisco Bay Bridge", Report UCB /EERC-93/12, University of California, Earthquake Engineering Research Center, Berkeley, CA.

60. J. E. y. C. A. Carballo, Cornell, 2000Probabilistic seismic demand analysis: spectrum matching and design", Department of Civil and Environmental Engineering, Stanford University, Report RMS-41

61. C. Kircher, 1993Personal communication with Farzad Naeim and Marshall Lew.

62. F. Naeim, J. Kelly, 1999Design of Seismic Isolated Structures from Theory to Practice. New York, John Wiley & Sons. 289p.

63. N. E. Huang, S. Zheng, S. R. Long, M. C. Wu, H. H. Shih, Q. Zheng, N. Yen, C. , C. C. Tung, M. H. Liu, 1998The empirical mode decomposition and Hilbert spectrum for nonlinear and nonstationary time series analysis", Proc. R. Soc. London, Ser. A 454903995

64. A. Roulle, F. J. Chavez-Garcia, 2006The strong ground motion in Mexico City: Analysis of data recorded by a 3D array, Soil. Dyn. Eq. Eng. 2671

65. H. Kawase, K. Aki, 1989A study of the response of a soft bas in for incident S, P and Rayleigh waves with special reference to the long duration observed in Mexico City, Bull Seism. Soc. Am. 7913611382

66. N. Yoshida, S. Iai, 1998Nonlinear site response and its evaluation and prediction," IN Irikura, K., Kudo, K., Okada, K. & Sasatani, T. (Eds.) The Second International Symposium on the Effects of Surface Geology on Seismic Motion, Yokohama, Japan, A.A.Balkema, 7190

67. I. y. Herrera, E. . Rosenblueth, Response Spectra on Stratified Soil". Proc. 3rd. World Conference on Earthquake Engineering. Nueva Zelandia, 44561965

68. M. P. Romo, A. Jaime, 1986Dynamic characteristics of some clays of the Mexico Valley and seismic response of the ground". Technical Report, Apr., Instituto de Ingenieria, Mexico City, Mexico (in Spanish).

69. A. Jaime, M. P. Romo, D. Reséndiz, 1988Comportamiento de pilotes de fricción en arcilla del valle de México." Series of the Instituto de Ingeniería, Mexico City, Mexico

70. M. P. Romo, Seed, 1986Analytical modelling of dynamic soil response in the Mexico Earthquake of September 19, 1985". Proc. ASCE Int. Conf. on the Mexico Earthquakes-1985148162

71. S. R. García, M. P. Romo, J. Mayoral, 2007Estimation of Peak Ground Accelerations for Mexican Subduction Zone Earthquakes using Neural Networks", Geofísica Internacional, 46-15163enero-marzo

72. G. Castro, S. J. Poulos, J. W. France, J. L. Enos, 1982Liquefaction induced by cyclic loading. Winchester, Mass: Geotechnical Engineers Inc.

73. H. B. Seed, I. M. Idriss, I. Arango, . Evaluation, Liquefaction. of, Using. Potential, Performance. Field, Data, of. Journal, Geotechnical. the, Division. A. S. C. E. Engineering, Vol, GT31983Seed et al., 1983

74. H. B. Seed, I. Idriss, M.1971Simplified procedure for evaluation soil liquefaction potential. Journal of the Soil Mechanics and Foundations ASCE, 97 (9), 1249-1273.

75. T. L. Youd, I. M. . Idriss, R. D. Andrus, I. Arango, G. Castro, J. T. Christian, R. Dobry, F. Liam, L. F. Harder, M. E. Hynes, K. Ishihara, J. P. Koester, S. S. C. Liao, I. I. I. W. F. Marcuson, G. R. Martin, J. K. Mitchell, Y. Moriwaki, M. S. Power, P. K. Robertson, R. B. Seed, K. H. Stokoe, 200, Liquefaction resistance of soils. Summary report from the 1996NCEER and 1998 NCEER/NSF workshops on evaluation of liquefaction resistance of soils. J. Geotech. Geoenviron. Eng., 127(10), 817-833.

76. R. Boulanger, I. M. Idriss, 2004State normalization of penetration resistance and the effect of overburden stress on liquefaction resistance. Proc. 11th International Conf. on Soil Dynamics and Earthquake Engineering and 3rd International Conference on Earthquake Geotechnical Engineering, Univ. of California, Berkeley, CA.

77. S. M. Fatemi-Agdha, M. Teshnehlab, A. Suzuki, T. Akiyoshi, Y. Kitazono, 1998Liquefaction potential assesment using multilayer artificial neural network. J. Sci. I.R. Iran, 9(3).

78. C. H. Juang, C. J. Chen, Y. M. Tien, 1999Appraising cone penetration test based liquefaction resistance evaluation methods: Artificial neural networks approach. Canadian Geotechnical Journal, 363443454

79. C. H. Juang, H. M. Yuan, D. H. Lee, Lin, 2003P. S., "Simplified cone penetration test-based method for evaluating liquefaction resistance of soils," Journal of Geotechnical and Geoenvironmental Engineering, 12916680

80. M. H. Baziar, N. Nilipour, 2003Evaluation of liquefaction potential using neural-networks and CPT results. Soil Dynamics and Earthquake Engineering, 237631636

81. S. Chern, C. Lee, 2009CPT-based simplified liquefaction assessment by using fuzzy-neural network. Journal of Marine Science and Technology, 174326331

82. García S.R. and Romo M.P.2007GENES: Genetic Generator of Signals, a Synthetic Accelerograms Application. , Proc. of the SEE5, SM-80.

83. S. F. Bartlett, T. L. Youd, 1992Empirical Analysis of Horizontal Ground Displacement Generated by Liquefaction Induced Lateral Spreads", Tech. Rept. NCEER 920021National Center for Earthquake Engineering Research, SUNY- Buffalo, Buffalo, NY.

84. Youd T.L., Hansen C.M. and Bartlett S.F.2002Revised Multilinear Regression Equations for Prediction of Lateral Spread Displacement' Journal of Geotechnical and Geoenvironmental Engineering, 1281210071017

CITATION

Silvia Garcia (2012). A Cognitive Look at Geotechnical Earthquake Engineering: Understanding the Multidimensionality of the Phenomena, Earthquake Engineering, Prof. Halil Sezen (Ed.), ISBN: 978-953-51-0694-4, InTech, DOI: 10.5772/50369.

CHAPTER 2

An Updated Seismic Source Model for Egypt

R. Sawires[1, 4,] J.A. Peláez[2], R.E. Fat-Helbary[3], H.A. Ibrahim[1] and M.T. García Hernández[2]

[1] Department of Geology, Faculty of Science, Assiut University, Assiut, Egypt
[2] Department of Physics, University of Jaén, Jaén, Spain
[3] Aswan Seismological Center, Aswan, Egypt
[4] Department of Physics, University of Jaén, Jaén, Spain

INTRODUCTION

Since the pioneering work of Cornell [1], it is clear that seismic hazard assessment depends on several models, among them perhaps one of the most significant, and usually poorly understood, is the delineation and characterization of the seismic source model for a particular region. Identification and characterization of the potential seismic sources in any region is one of the most important and critical inputs for doing seismic hazard analysis.

In fact, the characterization of seismic source zones depends on the interpretation of the available geological, geophysical and seismological data obtained by many tools such as tectonic studies, seismicity, surface geological investigations and subsurface geophysical techniques [2]. In addition, the characterization depends on the definition of different surface and sub-surface active faults.

Modern investigations on Probabilistic Seismic Hazard Assessment (PSHA) for any region at any scale, requires that the study region should be subdivided into different seismic sources. The issue of seismic source delineation and characterization is often a controversial one in the practice of seismic hazard analyses, both deterministic and probabilistic, as the information available relating to geology and seismotectonics can vary from region to another region.

It has been common practice since the development of PSHA by Cornell [1] and McGuire [3], to utilize areal source zones of seismic homogeneity [4 and 5]. In the classic form, earthquake sources range from clearly understood and well defined faults to less well understood and less well-defined geologic structures to hypothetical seismotectonic provinces extending over many thousands of square kilometers whose specific relationship to the earthquake generating process is not well known [2].

Recent PSHA at a local or a regional scale is usually based on approaches and computer codes (e.g., FRISK: [6]; SEISRISK III: [7]; CRISIS 2014: [8], etc.) that require the study area to be subdivided into seismic source zones which can be generated by delineating a number of polygons over active seismic areas. These polygons, sometimes have a complex shape, which reflects the complexity of the different faults and tectonic trends (e.g., [9]). The delineation will serve for two purposes: i) adequately represents the geological and tectonic setting together with the recorded seismicity, and ii) it allows for expected variations in future seismicity.

Seismicity and Seismotectonic Setting of Egypt

Egypt is situated in the northeastern corner of the African Plate, along the southeastern edge of the Eastern Mediterranean region. It is interacting with the Arabian and Eurasian Plates through divergent and convergent plate boundaries, respectively. Egypt is surrounded by three active tectonic plate boundaries: the African-Eurasian plate boundary, the Gulf of Suez-Red Sea plate boundary, and the Gulf of Aqaba-Dead Sea Transform Fault (Figure 1). The seismic activity of Egypt is due to the interaction and the relative motion between the plates of Eurasia, Africa and Arabia. Within the last decade, some areas in Egypt have been struck by significant earthquakes causing considerable damage. Such events were interpreted as the result of this interaction.

Based on the geophysical studies in the territory of Egypt, Youssef [10] classified the main structural elements of Egypt (Figure 2) into the following fault categories: a) Gulf of Suez-Red Sea, b) Gulf of Aqaba, c) east-west, d) north-south, and e) N45°W trends. However, Meshref [11], from the magnetic tectonic trend analysis, showed the tectonic trends which influenced Egypt throughout its geologic history as: a) NW (Rea Sea-Gulf of Suez), b) NNE (Aqaba), c) east–west (Tethyan or Mediterreanean Sea), d) north–south (Nubian or East African), e) WNW (Drag), f) ENE (Syrian Arc), and g) NE (Aualitic or Tibesti) trends.

The seismicity of Egypt has been studied by many authors [e.g., 12-22]. Although Egypt is an area of relatively low to moderate seismicity, it has experienced some damaging local shocks throughout its history, as well as the effects of larger earthquakes in the Hellenic Arc and the Eastern Mediterranean area. In addition, it has also been affected by earthquakes in Southern Palestine and the Northern Red Sea [18].

In Egypt, mostly population settlements are concentrated along the Nile Valley and Nile Delta, so, the seismic risk is generally related to the occurrence of moderate size earthquakes at short distances (e.g., M_S 5.9, 1992 Cairo earthquake), rather than bigger earthquakes that are known to occur at far distances along the Northern Red Sea, Gulf of Suez, and Gulf of Aqaba (e.g., M_S 6.9, 1969 Shedwan, and M_W 7.2, 1995 Gulf of Aqaba earthquakes), as well as the Mediterranean offshore (e.g., M_S 6.8, 1955 Alexandria earthquake) [23].

Egypt is suffering from both interplate and intraplate earthquakes; intraplate earthquakes are less frequent but still represent an important component of risk in Egypt. Shallow-depth seismicity (Figure 3) is concentrated mainly in the surrounding plate boundaries and on some active seismic zones like Aswan, Abu Dabbab, and Cairo-Suez regions, while the deeper activity is concentrated mainly along the Cyprian and Hellenic Arcs due to the subduction process between Africa and Europe.

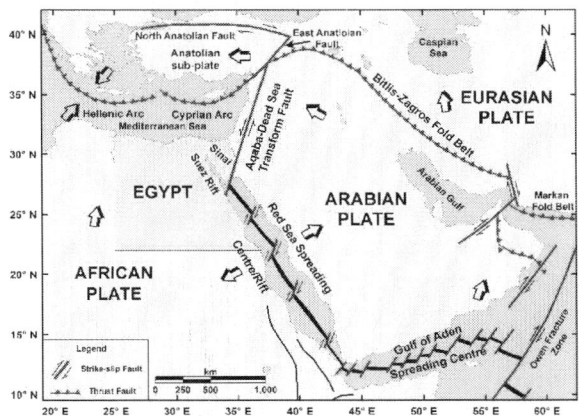

Figure 1: Global tectonic sketch for Egypt and its vicinity (redrawn after Ziegler [24] and Pollastro [25]).

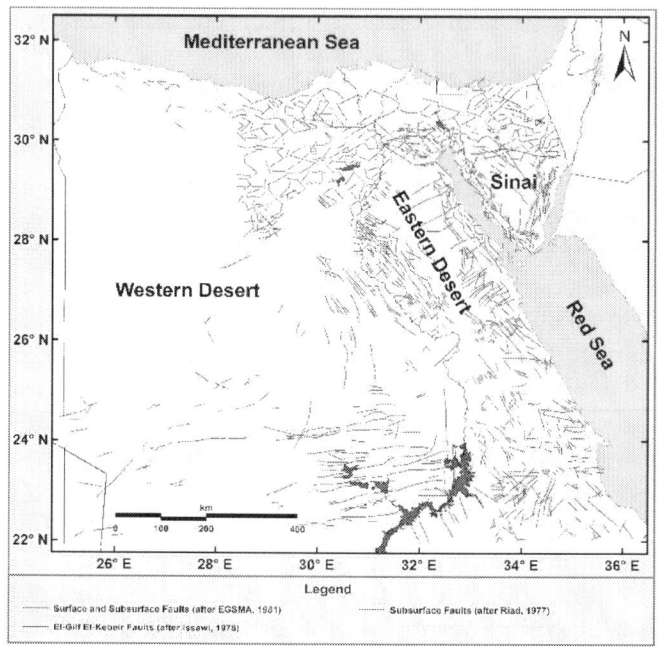

Figure 2: Distribution of major surface and subsurface faults. Compiled and redrawn from EGSMA [26] geologic map, from Riad [27], and from Issawi [28].

REVIEW OF SEISMIC ZONING STUDIES IN EGYPT

Seismic hazard assessments for Egypt, based on the zoning approach, has been carried out by many authors in the last decades, based upon the main tectonic features prevailed, the dominant tectonic stresses, the history of seismicity in the region, and the distribution of the recorded earthquakes. These authors were used different criteria to obtain seismic source zonation maps.

Among those studies, those carried out by the following authors: Sieberg [12 and 13], Gergawi and El-Khashab [15], Maamoun and Ibrahim [29], Maamoun *et al.* [16], Albert [30 and 31], Kebeasy *et al.* [32], Kebeasy [17 and 33], Marzouk [34], Fat-Helbary [35-37], Reborto *et al.* [38], Mohammed [39], El-Hadidy [40], Fat-Helbary and Ohta [41], El-Sayed and Wahlströrm [42], Abou Elenean [19 and 43], Badawy [44], Deif [45], Riad *et al.* [46], Abou Elenean and Deif [47], El-Sayed *et al.* [48], Fat-Helbary and Tealeb [49], El-Amin [50 and 51], El-Hefnawy *et al.* [52],

Abdel-Rahman *et al.* [53], El-Hadidy [54 and 55], Deif *et al.* [56 and 57], Fat-Helbary *et al.* [58] and Mohamed *et al.* [59].

Egypt was divided into different seismic zones by many researchers, using the distribution of historical and instrumental earthquakes. Maamoun and Ibrahim [29] and Kebeasy [33] divided Egypt into four main seismic trends: i) Northern Red Sea-Gulf of Suez-Cairo-Alexandria, ii) Eastern Mediterranean-Cairo-Fayoum, iii) Mediterranean Coastal Dislocation, and iv) Aqaba-Dead Sea Transform. More recently, Maamoun *et al.* [16] added another two trends to the previous four: i) Hellenic and Cyprian Arcs, and ii) Southern Egyptian trend.

In reviewing the seismicity of Egypt, Kebeasy [17] suggested three main seismic zones: i) Aqaba-Dead Sea Transform, ii) Northern Red Sea-Gulf of Suez-Cairo-Alexandria, iii) Eastern Mediterranean-Cairo-Fayoum zones. In addition, he defined other local seismic zones (e.g., El-Gilf El-Kebeir, Aswan and Qena zones).

Fat-Helbary [36] assessed the seismic hazard for Aswan region. He used both of line sources and area source models. Five active faults in the Aswan region (Kalabsha, Seiyal, Gebel El-Barqa, Kurkur, and Khur El-Ramla Faults) were modeled as seismic lines. On the other hand, six area source zones (Old Stream, North Kalabsha, Khur El-Ramla, East Gebel Marawa, Abu Dirwa, and Kalabsha zones) were considered in the assessment. This study was followed by successive assessments by different authors to include other neighbor regions in Upper Egypt (e.g., [37, 41, 49, 50, 51, 57 and 58]).

Using the relation between the paleo-stresses, the present-day stresses and the distribution of earthquake epicenters, El-Hadidy [40] deduced five major trends in Egypt. They are: i) Pelusium megashear, ii) Eastern Mediterranean-Cairo-Fayoum-El-Gilf El-Kebeir, iii) Nubian-Mozambique, iv) Qena-Aqaba-Dead Sea, and v) Northern Red Sea-Gulf of Suez-Cairo-Alexandria seismotectonic trends. Furthermore, he identified some local zones on the Red Sea, Gulf of Suez, Gulf of Aqaba, Nile Delta, and Cairo-Suez regions.

According to the earthquake distribution, focal mechanisms and the structural and tectonic information, Abou Elenean [19] suggested five seismotectonic sources. They are: i) Gulf of Suez-Northern Eastern Desert, ii) Southwest Cairo (Dahshour), iii) Northern Red Sea, iv) Gulf of Aqaba, and v) Aswan zones. Deif [45], for a seismic hazard assessment study, delineated four additional seismic sources for the southern part of Egypt. They are: i) Abu Dabbab, ii) El-Gilf El-Kebeir, iii) Wadi Halfa, and iv) Northern Nasser's Lake zones.

Riad *et al.* [46] constructed a more detailed seismic zoning map for Egypt and its surroundings. Their regional delineation consists of five main trends: i) the Greek trend, based on the seismic zone regionalization of Papazachos [61], ii) the Dead Sea trend, which mainly based on the earthquake catalogue of Israel and its vicinity [62], iii) Pelusium and Qattara trend, iv) Eastern Mediterranean trend, and v) Aswan area, in Southern Egypt.

El-Hefnawy *et al.* [52], based on the tectonic regime, seismicity, faults location, and focal mechanism solutions, divided the regional seismicity in and around Sinai Peninsula into 25 source zones. His study was succeeded by a certain number of studies that considered a more detailed zonation for the same area (e.g., [53, 54 and 56]).

Recently, Abou Elenean [43] established a detailed zonation map for whole Egypt and its surroundings, considering the recent seismicity distribution and focal mechanism data. He delineated 41 seismic source zones of shallow-depth earthquakes (h < 60 km) in and around Egypt. In addition, he considered 7 seismic sources for intermediate-depth events within the Hellenic Arc (after [63]). More recently, El-Hadidy [55] and Mohamed *et al.* [59] established a new and modified seismic zoning map for Egypt and its surroundings which is based on the compilation of previous studies [53, 57 and 64].

DATA SOURCES

For the construction of any database of seismic sources, there are two basic steps: first, all of the active faults that affect a specific region need to be recognized, and secondly, each seismogenic structure should be seismotectonically parameterized. In order to recognize the active faults, it is necessary to analyze the seismicity. It is common practice to start analyzing the historical and instrumental seismicity that affects the specific region. Like many other places all over the world, the seismicity in Egypt is not homogeneously distributed, neither in frequency nor in density. Historical information is similarly not uniform all over the region.

An Updated Earthquake Catalogue

A complete and consistent earthquake catalogue in a region is essential in order to study the distribution of earthquakes in space, time, and magnitude. In the current work, the identification and characterization of regional seismic source zones is based on a unified compiled earthquake catalogue, after Sawires *et al.* [60], for Egypt and its surroundings which covers the area from 21° to 38° N and 22° to 38° E, and extends from 2200 B.C. until 2013 in the time period.

Different earthquake magnitude scaling relations, correlating different scale magnitudes, were used to develop a unified earthquake catalogue for the study region in the moment magnitude (M_W) scale. The dependent events were removed from the catalogue to ensure a time-independent (Poissonian) distribution of earthquakes (Figure 3).

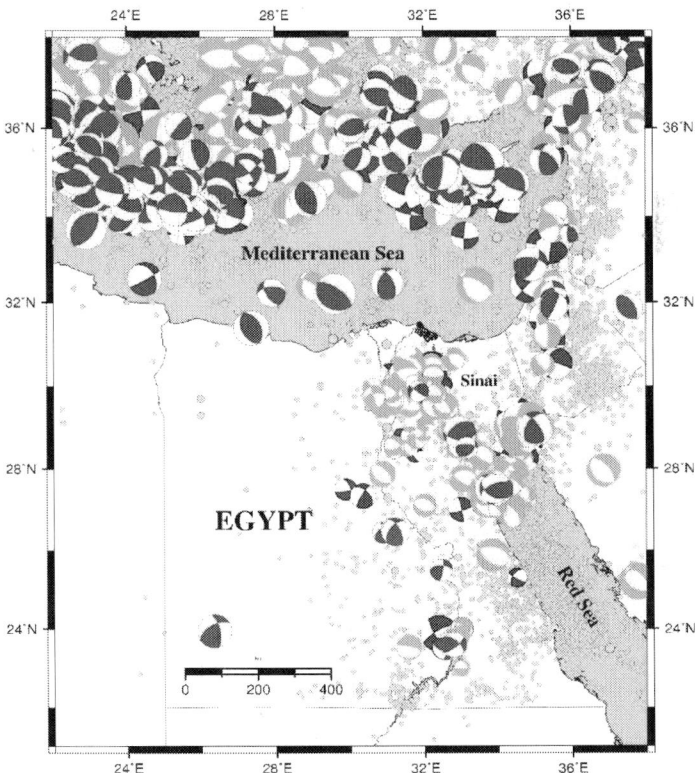

Figure 3: Distribution of the seismicity (2200 B.C. - 2013) and focal mechanism solutions (1940 – 2013) in and around Egypt (after Sawires *et al.* [60]). Symbols and focal sphere sizes are in proportion the moment magnitude. Focal sphere colours refer to different fault types (blue: strike-slip; green: normal; red: reverse)

Focal Mechanism Data

Different local and international sources were examined and focal mechanism data were compiled into a single database. The solutions of the Global Catalogue of CMT Harvard [65], the International Seismological Centre (ISC) [66], the National Earthquake Information Centre (NEIC) [67], the Regional CMT catalogues (RCMT) in the Mediterranean region [68], as well as ZUR-RMT catalogue of the Institute of Technology (ETH) of Zurich were also included in the catalogue. More than 600 focal mechanism solutions were collected covering different active seismic zones (Figure 3) in Egypt and surroundings, spanning the spatial area from 21° to 38°N, and from 22° to 38°E. Most of them have a magnitude greater than or equal to M_W 3.0, occurring in the time period 1940 to 2013.

Geological, Tectonic and Geophysical Data
Several geological, geophysical and tectonic maps were inspected for the purpose of getting more information about the present active faults (e.g., Aswan region) and also for the identification of the prevailed tectonic and structural trends in the study region. Among these studies are those of Said [69-71], Youssef [10], Shata [72], Neev [73], Neev et al. [74 and 75], El-Shazly [76], Riad [27], Maamoun [77], Issawi [78], EGSMA [26], Riad et al. [47 and 79], Maamoun et al. [16], Sestini [80], Schlumberger [81], Woodward-Clyde Consultants [82], Kebeasy [17], Meshref [11], Barazangi et al.[83], Guiraud and Bosworth [84], Abdel Aal et al. [85], Philobbos et al. [86], and Hussein and Abdallah [87].

Crustal Structure Data
The crustal structure plays an important role in Seismology. It can be used, as in the current study, for the discrimination between the crustal (shallow-depth) seismicity, the intermediate-depth, and the deeper one.

Several studies have been carried out to evaluate the crustal structure and thickness in Egypt by using different types of datasets coming from seismic reflection surveys, deep seismic sounding, shallow refractions, and gravity (e.g., [34, 40 and 88-108]). In the delimitation of the different seismic zones, the most recent study [108] was taken into our consideration (Figure 4). Their results show that the Moho discontinuity is getting shallow toward the northern and eastern coast of Egypt, and deeper toward Western Desert and Northeastern Sinai. This discontinuity is located at depth of 31-33 km in Greater Cairo and Dahshour, 32-35 km in Sinai, 33–35 km along the Nile River, 30 km near the Red Sea coast, and 39 km towards the Western Desert.

DETAILED DESCRIPTION OF THE NEW PROPOSED SHALLOW-DEPTH SEISMIC SOURCE MODEL

Seismic sources define areas that share common seismological, tectonic, and geologic attributes, and that can be described by a unique magnitude-frequency relation. In terms of PSHA, a seismic source represents a region of the earth's crust in which future seismicity is assumed to follow specified probability distributions for occurrences in time, earthquake sizes, and locations in space [109].

Araya and Der Kiureghian [109] discriminate between seismogenic and seismicity sources. Seismogenic zones lack the development of a clear history relating the contemporary seismic activity to a geologic structure. For such zones, critical gaps in the Quaternary geologic history preclude direct evidence of active faulting. Seismogenic zones are, by far, the most common type of source zone employed in PSHA. Commonly, seismogenic zones are area sources, but the zone type applies also to inferred associations of seismicity with individual faults. On the other hand, seismicity zones are source zones that are defined with no consideration of their relation to geologic structures. They are defined solely based on the spatial distributions of the seismic history, and their use and reasonableness can only be judged relative to the intended use of the final hazard estimate. This will be the terminology used in this work.

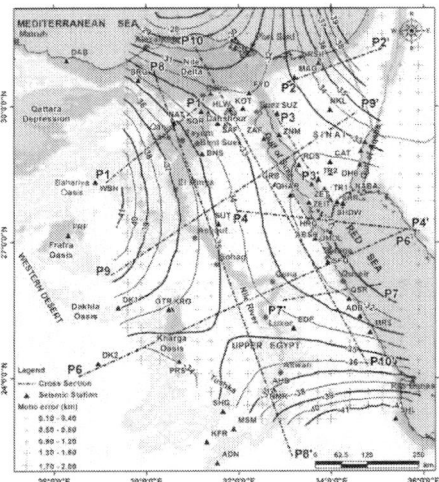

Figure 4: Depth of Moho discontinuity in Egypt (after Abdelwahed *et al.* [108]).

As mentioned previously, the separation of the study area into smaller, seismotectonically homogeneous zones is based on criteria mainly related with the present-day tectonic regime, epicenter distribution, focal mechanism data and the location of known faults. In the present work, we decided to employ simple geometric shapes for the definition of the seismic source model. The regional seismicity of concern to Egypt was divided into 28 seismic sources (Figure 5). These zones was related to the tectonic activity of the previously defined local active belts. Thus, the majority of the proposed sources zones can be considered seismogenic zones, except some sources which can be considered seismicity sources. The delineation of the seismicity sources was based upon the earthquake distribution, this is because there is no enough geologic and tectonic data covering these sources. Both seismogenic and seismicity sources are described below in more details. For each of these source zones, the seismicity parameters (b-value and activity rates) were computed by applying the Gutenberg-Richter [110] relationship and using the least square method considering the entire earthquake events within each zone. Moreover, maximum observed magnitude M_{max} was defined using the earthquake sub-catalogue for each source. Those estimated values will serve as initial inputs for a seismic hazard assessment for Egypt in the near future.

The details of the selection of these seismic sources, together with the estimation of its seismicity parameters and maximum observed magnitude, are given below for each source category, which grouped depending on the similarities of the prevailed tectonic environment.

Seismic Sources Along The Gulf Of Aqaba–Dead Sea Transform Fault

The Aqaba-Dead Sea Transform Fault (DST) is a 1100 km long left-lateral strike-slip fault (Figure 6)that accommodates the relative motion between Africa and Arabia [111, 112]. It is a seismically active transform boundary, connecting the Red Sea spreading center in the south to the Northern Mediterranean Triple Junction to the north. Its main left-lateral sense of motion is recognized by minor pull-aparts in young sediments [113], cut and offset of drainage lines and man-made structures (e.g., [114-121]).

The Gulf of Aqaba-Dead Sea Transform Fault (Figure 6) is subdivided into three parts; southern, central and northern [122]. The first part, which starts from the Gulf of Aqaba and passing through the Dead Sea and the Jordan Valley, is characterized by the occurrence of N12°E to N20°E left-lateral strike-slip faults. The second part of the DST is characterized by the occurrence of about 200 km long NNE–SSW restraining bend, where the

DST branches into different faults. The major one, called the Yammouneh Fault, which connects the first and third parts of the DST, while the other faults connect the DST with the Palmyride Fold Belt (PFB) [122-124]. The last and the northern part of the DST is characterized by the occurrence of two different N–S striking faults surrounding the Ghab Valley and intersecting through a complex braided fault system with the East Anatolian Fault and the Cyprian Arc [125-127]. This intersection corresponds to the Hatay "fault–fault–trench" triple junction that forms the plate boundaries between Arabia, Africa and Anatolia [128].

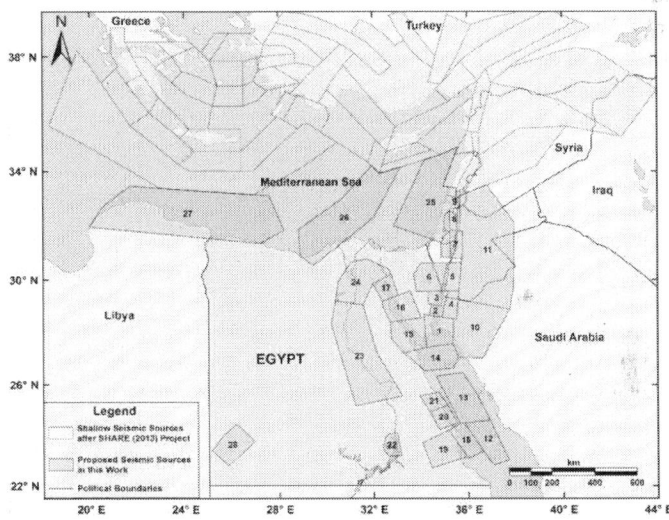

Figure 5: Proposed seismic source zones in Egypt and its surroundings.

Gulf Of AQABA Seismogenic Sources (Eg-01 Till Eg-04)
The Gulf of Aqaba experienced the largest Egyptian earthquake (M_W 7.2, November 1995) which struck the area and its effects were extending till Cairo. Over than 1000 aftershocks are recorded. The aftershocks area reached a length of about 110 km, striking N 30° E, which in turn parallel to the Gulf of Aqaba trend [129]. Potential damage was observed at Nuweiba city at the western part of the gulf.

The Gulf of Aqaba has been considered to be the most active seismic area over the last few decades, characterized by swarm activity [130-132]. There is no information about the seismicity of the Gulf of Aqaba until the year 1983. However, from January till April 1983, over than 500 events were reported, reaching a maximum recorded magnitude of 4.8. These earthquake events were felt at different places along the gulf area, as well as along the Arava Valley founding a general consideration [133]. From August 1993 up to February 1994, a large earthquake swarm was associated with relatively high magnitudes, reaching a 5.8 value. This swarm included about 1200 events occurred south to the 1983 swarm. Another earthquake swarm has been recorded and located at the central part of the Gulf of Aqaba on November 2002. Over than 10 events with magnitude above 4.0 were recognized, and many other events with magnitudes below this value. Some of these earthquakes were felt, but without damage for buildings at the epicentral area.

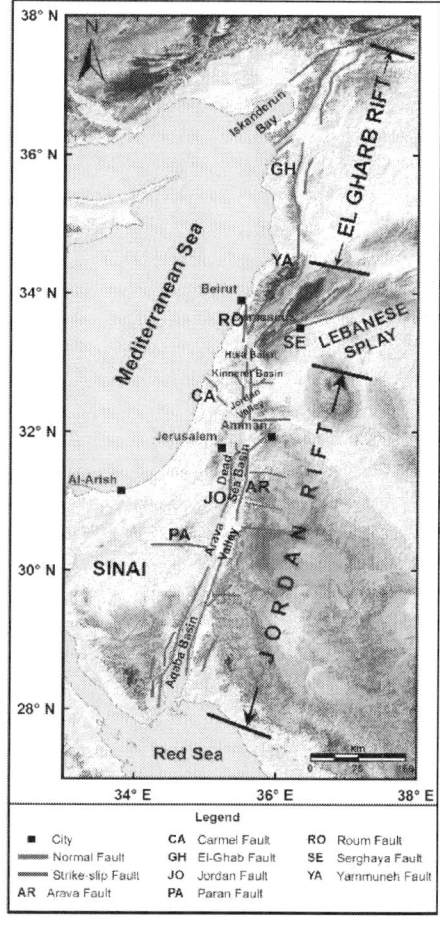

Figure 6: The main structural elements along the DST (redrawn after Heidbach and Ben-Avraham [134]).

The interior of the Gulf of Aqaba is occupied by three elongated en-echelon basins transected by longitudinal faults [131]. This en-echelon system produces several tectonic basins, which are forming rhombic-shaped grabens. Thus, three basins in the Gulf of Aqaba are present. They are, from south to north, Tiran "Arnona"-Dakar, Aragonese and Elat "Aqaba" Basins.

The heterogeneity of the focal mechanism solutions for the earthquake events taken place in the gulf area, indicates its geologic structure complexity. Some fault plane solutions exhibit normal faulting, which are related to the faults that form the boundaries of the major basins in the gulf. Others indicate left-lateral motion of the transform [112]. The focal mechanism of the M_W 7.2, 1995 Aqaba earthquake as well as some aftershocks, show a strike-slip movement with predominant normal components, with the exception of only one solution located on the eastern coast of the Gulf of Aqaba, and exhibits strike-slip movement with a little reverse component in the NNW-SSE and ENE-WNW nodal planes [19]. According to the seismic activity, the epicentral distribution and the local tectonics, different seismogenic sources were delineated in the gulf area (Figure 7).

a. The EG-01 (Tiran – Dakar Basin) seismogenic source lies at the southern part of the Gulf of Aqaba. It includes the M_S 4.4, February 2, 2006 earthquake. There is no historical earthquakes included in this source zone. The majority of the available focal mechanism solutions inside this area source reflects normal faulting mechanism.

b. The EG-02 (Aragonese Basin) seismogenic source lies to the north of the previous EG-01 zone, and is considered the focal area of the M_W 7.2, November 22, 1995 earthquake, which is considered the largest event to occur along the DST in the last century.

c. The EG-03 (Elat Basin) seismogenic source located to the north of the EG-02 seismic zone and considered as the extension area of the M_W 7.2, 1995 Aqaba earthquake rupture. It is characterized by a low seismicity level, if compared with the other two zones of the Gulf of Aqaba. Two historical events have been included in this area source, the I_{max} VIII, March 18, 1068, and the I_{max} VIII-IX, May 2, 1212 earthquakes.

d. In addition to the previous seismogenic sources, a delineation of a separate and fourth zone is taken place. This source lies to the east of the gulf and characterized by dispersed moderate seismicity. This zone is the EG-04 (Eastern Gulf of Aqaba) seismogenic source. The major earthquake included in this area source is the m_b 4.5 December 26, 1995 earthquake.

Previous focal mechanism solution studies for moderate to large earthquakes located in the Gulf of Aqaba region (e.g., [135-138]) assert the dominance of ENE-WSW extension (N60°- 80°E). Furthermore, field studies [139, 140] observed two conjugate faults along the Gulf of Aqaba: NNE left-lateral strike-slip faults parallel to the gulf that release the majority of stress, and a nearly ESE-WNW normal faults along the margins of pull-apart basins. On the other hand, body waveform inversion of the M_W 6.1, August 3, 1993, and the M_W 7.2 November 22, 1995 events, support the occurrence of normal faulting take place along the transverse NNW-SSE and ESE-WNW faults, while left-lateral strike-slip movement occurs along NNE major Aqaba trend [135].

Arava Valley (Eg-05) Seismogenic Source

The Arava Valley is located to the north of the Gulf of Aqaba. It is an inter-basin zone trending NE-SW. Its faults extend over 160 km from the Gulf of Aqaba to the Dead Sea and provide morphological evidence of essentially strike-slip motion [120]. It is characterized by a low seismicity level compared with the surrounding area, despite clear indications of recent faulting [141]. Klinger *et al.* [120] emphasized the limited earthquake activity in the Arava Valley in the instrumental period. Shapira and Jarradat [133] stated that, from preliminary paleoseismicity studies, the border-faults of Arava Valley generate earthquakes bigger than magnitude 6.0 with an average return period of 1000-3000 years.

There is no historical earthquakes included in this seismogenic source zone. The biggest recorded event is the m_b 5.2, December 18, 1956 earthquake. Two focal mechanism solutions are known in the northern part of this source, both of them exhibiting strike-slip faulting with normal component.

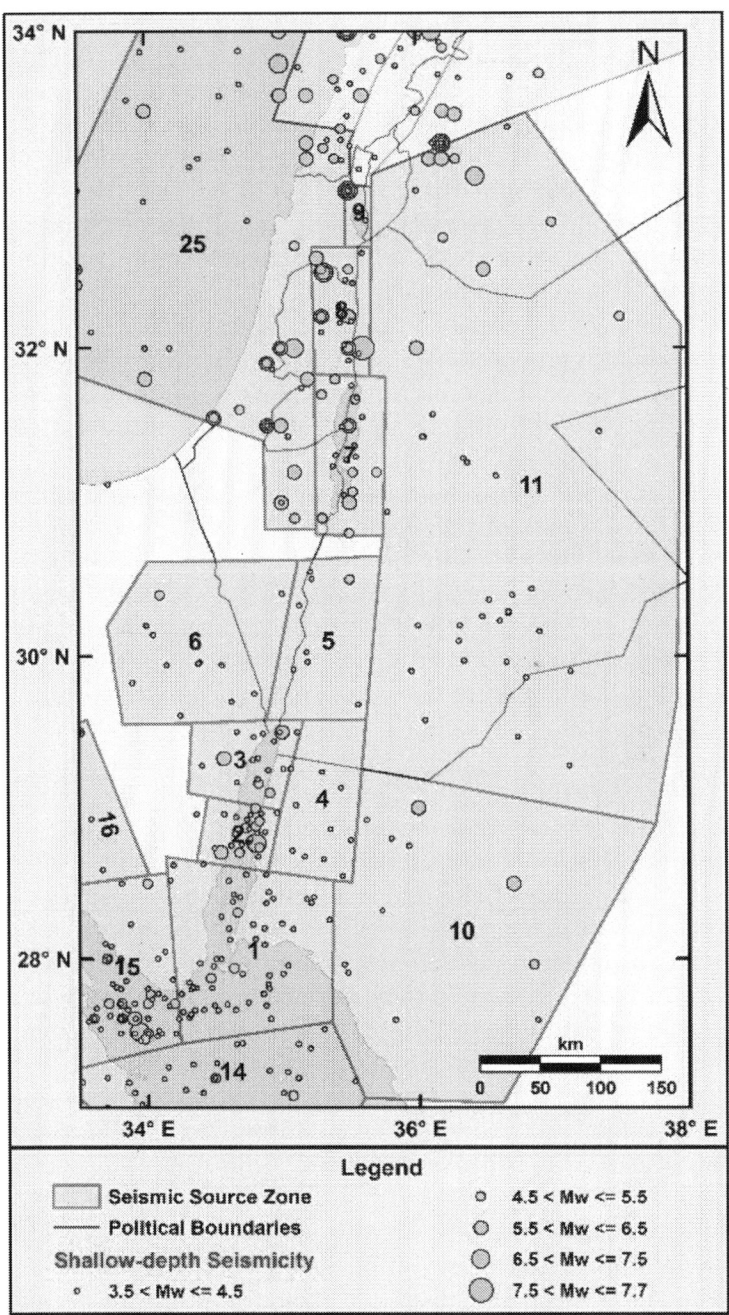

Figure 7: Shallow-depth seismicity (h ≤ 35 km) and delineated seismic sources along the Gulf of Aqaba-Dead Sea Transform Fault.

Eastern Central Sinai (Eg-06) Seismogenic Source

An E-W trending dextral strike-slip faults with up to 2.5 km of displacement has been recognized in central Sinai by Steintz *et al.* [142]. It is called the Themed Fault. The Tih Plateau (in central Sinai) is traversed by the Themed Fault, which extends for about 200 km from the vicinity of eastern margin of the Suez Rift to the DST [71]. The Themed Fault has been reactivated along a pre-existing fault, identifying the southern border of the Early Mesozoic passive continental margin of the Eastern Mediterranean Basin in central Sinai [143].

To the north of the previous fault, the central Sinai-Negav shear zone is located, which is proposed by Shata [72] and Bartov [144]. It is a narrow E to ENE trending fault belt discriminating and separating the North Sinai Fold Belt (tectonically unstable area) from the Tih Plateau (tectonically stable area) in middle and Southern Sinai [145].

The EG-06 seismogenic source lies to the west of the previous EG-05 source and to the east of the Sinai sub-Plate. This seismogenic source includes the low seismic activity related to the Themed Fault, central Sinai-Negav shear zone, Paran Fault and Baraq/Paran Fault junction. This source has a great tectonic effect on Sinai Peninsula and its surrounding areas. There is no historical earthquakes included in this source, and the biggest earthquake located in this zone is the m_b 4.8, September 24, 1927 event.

Dead Sea Basin (Eg-07) Seismogenic Source

The Dead Sea Basin is characterized by a double fault system that is bounded by the Arava Fault from the east, and by the Jordan (Jericho) Fault from the west, hence it occupies a rhomb-shaped graben between two left-lateral slip faults. The average slip rate on the Dead Sea portion of the transform fault is estimated to be 0.7 cm/yr. [114], which is consistent with the average slip of the overall plate boundary of 0.7-1.0 cm/yr.

Earthquake swarms and a mainshock-aftershock type of activity characterize this seismogenic source. Trenching studies across the Jordan Fault indicate that two large earthquake swarms occurred since about 2000 years ago. One of them is between 200 B.C.- 200 A.D., while the other one is between 700 A.D.- 900 A.D. [114]. El-Isa *et al.* [146] attributes these swarms to subsurface magmatic activity and/or to the isostatic adjustments along the Gulf of Aqaba.

Several historical earthquakes are included inside this source zone. They are the 745 B.C., 33 A.D., 1048, 1212, 1293, and 1458 earthquakes. Their intensities range between VII to VIII. Ben-Menahem*et al.* [147] obtained focal mechanism solutions for some recent events (e.g., the M_S 4.9, October 8, 1970 earthquake) which took place in the Dead Sea area. All solutions indicate a left-lateral strike-slip movement on a sub-vertical fault striking with an average trend of N8°-10°E. However, Salamon et al. [112] obtained normal focal mechanism solutions for some relatively recent events. These solutions may be describe the earthquake activity of the N-S striking normal faults bordering the Dead Sea Basin. Field observations confirmed this type of activity [113, 148].

Jordan Valley (Eg-08) Seismogenic Source

The Jordan Valley trends in the N-S direction, linking between the Hula Basin to the north and the Dead Sea Basin to the south. The details about its end in the Sea of Galilee are not clear from the surface features [147]. Garfunkel et al. [113] noticed a small amount of compression along the valley and near the Jordan Fault trace. Recent earthquake activity along the Jordan Valley is low compared to the Southern Dead Sea Basin. Ten historical events (before 1900) are included in this area source. They are the 1020 B.C., 578 A.D., 580, 746, 854, 1034, 1105, 1160, 1260, and 1287 events. Their intensities range between IV to XI. The most important earthquake included in this source zone is the I_{max} XI, 746 event.

Kineret-Hula Basin (Eg-09) Seismogenic Source

To the north of the previous Jordan Valley source are located the Hula (Shamir-Almagor Fault) and Kineret (Kineret-Sheikh Ali Fault) Basins [149]. Seismic activity in the two mentioned basins was located till the Yammuneh Fault (NE-bend of the Dead Sea Transform). This area source, which surrounded by the Roum Fault from the western side and the Jordan Fault from the eastern side, was considered by Shamir et al. [150] as a seismogenic step zone. Three historical events are included in this zone. They are the 19 A.D., 419, and 756 earthquakes. The biggest earthquake is the I_{max} X, 19 A.D. event.

Northwestern Saudi Arabia (Eg-10) Seismicity Source

To the east of the EG-01 and EG-02 seismogenic sources, the Northwestern Saudi Arabia EG-10 source has been considered. This source zone covers disperse, low seismicity in the northwestern part of Saudi Arabia. Two historical events are reported to occur inside this area source. They are the March 18, 1068, and January 4, 1588 earthquakes, both of them with intensity VIII.

Lebanon (Eg-11) Seismicity Source

To the north of the previous seismic source and along the eastern boundaries of the EG-04, EG-05, EG-07, EG-08, and EG-09 sources, the Lebanon EG-11 seismicity source has been considered. This area source covers a dense disperse low-magnitude seismicity in Lebanon and Southern Syria. Nine historical earthquakes are located inside this area source. The most important among them are the 972, 1159, and 1182 events. Their felt intensities are I_{max} IX, IX-X, and IX, respectively.

The computed b-value, the annual rate of earthquakes, and the observed recorded maximum magnitude for the delineated seismic sources along the Gulf of Aqaba-Dead Sea Transform Fault are displayed inTable 1.

Table 1: b-value, annual rate of earthquakes, and maximum observed magnitude for the delineated seismic source zones along the Gulf of Aqaba-Dead Sea Transform Fault.

Source Zone	b-value	Yearly Number of Earthquakes Above $M_W 4.0$	Above $M_W 5.0$	Observed M_{max}
EG-01	1.13	0.9799	0.0732	m_b 4.4 on 2006/02/02
EG-02	0.98	0.4952	0.0521	M_W 7.2 on 1995/11/22
EG-03	0.97	0.2763	0.0296	I_{max} VIII-IX on 1212/05/02
EG-04	1.01	0.1961	0.0191	m_b 4.5 on 1995/12/26
EG-05	0.88	0.1882	0.0251	m_b 5.2 on 1956/12/18
EG-06	1.12	0.1853	0.014	m_b 4.8 on 1927/09/24
EG-07	0.87	0.3232	0.0438	I_{max} VIII on 1458/11/12*
EG-08	0.71	0.1865	0.0366	I_{max} XI on 0746/--/--
EG-09	0.91	0.0651	0.008	I_{max} X on 0019/--/--
EG-10	1.03	0.1934	0.018	I_{max} VIII on 1588/01/04*
EG-11	0.97	0.3645	0.0388	I_{max} IX-X on 1159/06/06

(*) the most recent event.

Seismic Sources Along The Red Sea Rift

The Arabian Plate is continuing to rotate away from the African Plate along the Red Sea Rift spreading center. The Red Sea occupies a long and slightly sinuous NW-trending escarpment-bound basin, 250-450 km wide and 1900 km long, between the uplifted shoulders of the African and Arabian shields. It is part of a rift system extending from the Gulf of Aden to the northern end of the Gulf of Suez. The overall trend of the rift is N30°W, although a few kinks occur at around 15°N, 18°N, and 22°N.

Depending on the structural setting and morphology of the Red Sea, it can be subdivided into four different zones (Figure 8). Each zone are representing distinct stage in the development of the continental margin and the generation of the mid-ocean ridge spreading system [151, 152]. These zones are:

i. *Active sea-floor spreading (Southern Red Sea):* It is located between 15°N and 20°N and characterized by a well-developed axial trough which has developed through normal sea-floor spreading during the last 5 Ma [153-155] or even older, at about 9–12 Ma [156].

ii. *Transition zone (central Red Sea):* It is located between 20°N to about 23°20′N, where the axial trough becomes discontinuous, in which the central Red Sea consists of a series of 'deeps' alternating with shallow 'inter-trough zones' [157]. An identical zone may flanks the deep axial trough between the side walls of the shallow main trough on both sides of other zones [151, 152].

iii. *Late stage continental rifting (Northern Red Sea):* This zone composed of a wide trough without a distinct spreading center, in spite of a number of small isolated "deeps" is occurred [152].

iv. *Active rifting:* This zone representing the expected line along which the Southern Red Sea may be propagate through the Danakil Depression Afar. This zone may be considered separately or it can be added to the first mentioned zone.

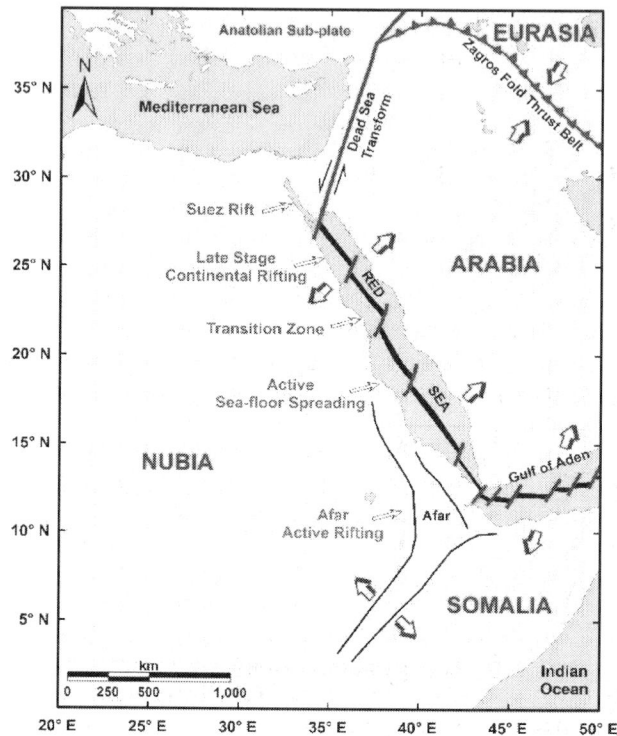

Figure 8: Tectonic framework of the Red Sea region (redrawn after Ghebreab [158]).

Based on the morphological and structural features of the Red Sea, the Egyptian part (northern latitude 22°N) can be divided into three distinct seismogenic source zones (EG-12, EG-13, and EG-14) (Figure 9). Each zone represents different stage of development [159]. The delineation is made, based upon the occurrence of the transverse structures, change of the fault trend along the axial rift and the variety of the seismic activity along the rift axis.

Southern Red Sea (Eg-12) Seismogenic Source

The EG-12 Southern Egyptian Red Sea seismogenic source represents the northern part of the transition zone. It is characterized by NW-SE trending faults. The boundary proposed by Bonati [160], north latitude 25°N, is found herein to coincide with the NE-trending transform faults and the associated seismicity. Only one historical event is included in this seismic source, the I_{max} VI-VII, 1121 earthquake.

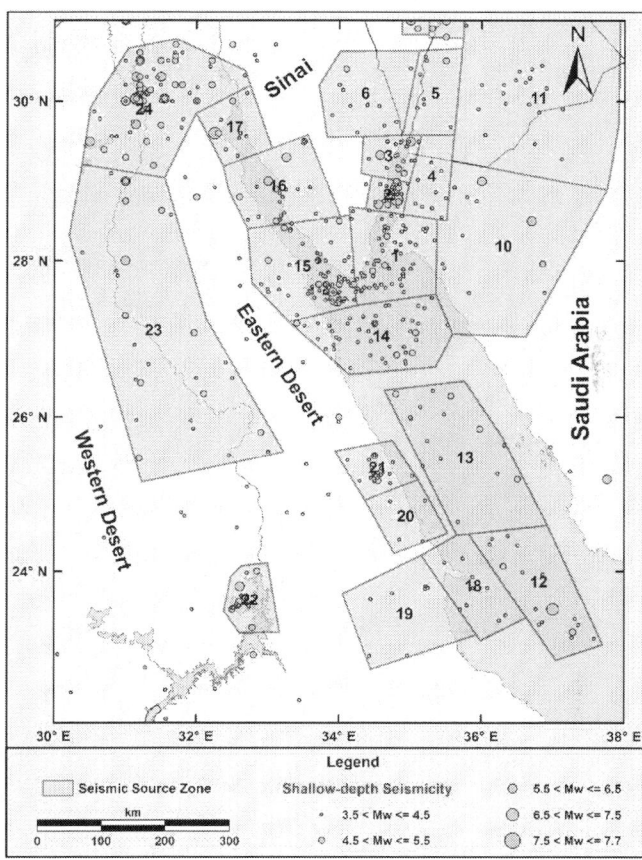

Figure 9: Shallow-depth seismicity (h ≤ 35 km) and delineated seismic sources along the Red Sea-Gulf of Suez and the Nile River.

Central Red Sea (Eg-13) Seismogenic Source

The EG-13 Central Egyptian Red Sea seismogenic source is located to the northwest of the previous zone. It corresponds to the region north of latitude 24°30′N, which consists of a broad main trough without a recognizable spreading center [152]. Recent recorded seismicity could indicate the expected location of the axial rift. In this zone, the degree of seismicity is relatively low and scattered, compared to the previous zone. Like the previous zone, there is only one historical event included here. It is the I_{max} V, 1899 earthquake. The maximum observed magnitude along this source corresponds to the m_b 4.7 (M_S 5.1), July 30, 2006 earthquake.

Northern Red Sea (Eg-14) Seismogenic Source

The EG-14 Northern Egyptian Red Sea seismogenic source is characterized by higher seismic activity than the previous two sources. This activity may be due to the juncture between the two gulfs. Daggett*et al.* [161] studies of the low-magnitude seismicity shows that, the high seismic activity of the northern Red Sea is different from the activity at the southern part of the Gulf of Suez. There is no earthquakes related to this area source before the year 1900. In addition, the m_b 5.0 (M_S 5.0) March 22, 1952 event represents the biggest recorded earthquake till now.

Seismicity parameters for the delineated seismic sources along the Red Sea Rift are displayed in Table 2.

Table 2: b-value, annual rate of earthquakes, and maximum observed magnitude for the delineated seismic source zones along the Red Sea Rift.

Source Zone	b-value	Yearly Number of Earthquakes		Observed M_{max}
		Above $M_W 4.0$	Above $M_W 5.0$	
EG-12	1	0.4359	0.0434	I_{max} VI-VII on 1121/- -/--
EG-13	0.91	0.3029	0.0376	m_b 4.7 on 2006/07/30
EG-14	1.13	0.6425	0.0472	m_b 5.0 on 1952/03/22

Seismic Sources Along The Gulf Of Suez

The Gulf of Suez is considered to be the plate boundary between the African Plate and Sinai sub-Plate [162]. It extends along a NW trend from latitude 27°30′ N to 30°N. The Gulf of Suez constitutes the northern part of the Red Sea Rift System. It was developed, together with the Red Sea and the Gulf of Aqaba, as one of the three arms of the Sinai Triple Junction [69, 81 and 163-166].

The Gulf of Suez has been interpreted as being a complex half-graben system [139], or an asymmetric graben [167]. It is composed of three successive half-grabens, as mentioned by Moustafa [168], with opposite tilt directions: northern, central, and southern. These distinct half-grabens include several rift blocks of a uniform dip direction. The dip direction, along the Gulf of Suez Rift, changes from the north to the south as: SW to NE and again to SW defining the three half-grabens, respectively.

Two-accommodation zones [169] coexist among these half-grabens which extend transversely across the rift (Figure 10). These are the Galala-

Zenima [168] or Gharandal [167] accommodation zone, of broad extension (about 60 km wide) in the north, and the Morgan [168] or Sufr El Dara [170] accommodation zone (20 km wide) in the south. Both zones exhibit a broad range of deformation, including distinct normal, oblique, or strike-slip faults [171], or wide complex zones of normal faulting, trans-tension [172-174] or broad warping [175].

The Gulf of Suez is considered to be an aseismic area during the first half of the last century and this consideration let some researchers (e.g., [176, 177]) to conclude that all the present motion taking place in the Red Sea Rift is transferred into shearing along the DST. Ben-Menahem [178] and Salamon*et al.* [111] studied the seismic activity of the Suez Rift. Fault plane solutions of the m_b 6.1, March 31, 1969 earthquake and other low-magnitude events show that the Gulf of Suez Rift is active which agree with Ben-Menahem and Aboodi [179] results. Considering the tectonic setting, seismicity and earthquake faulting mechanisms, the Gulf of Suez can be divided into three seismogenic sources (Figure 9) as follow.

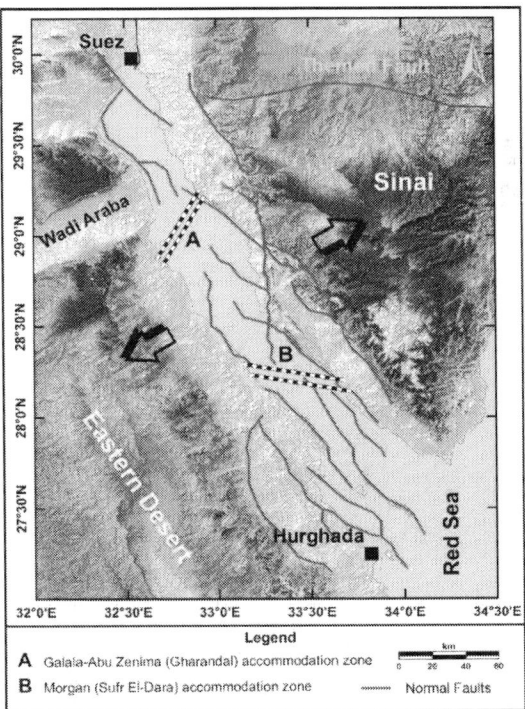

Figure 10: Tectonic setting of the Gulf of Suez. Red lines refer to normal faults (redrawn after Meshref [11]; and Younes and McClay [180]).

Southern Gulf Of Suez (Eg-15) Seismogenic Source

The EG-15 Southern Gulf of Suez seismogenic source is distinguished by intensive structural deformation. It is characterized by its relatively high seismic activity. The higher seismicity rate at the southern part of the Gulf of Suez is related to the crustal movements among the three surrounding plates: Arabian Plate, African Plate, and Sinai sub-Plate. Six historical events are included in this zone. Those are 28 B.C., 955, 1091, 1195, 1778, and 1839 events. Their intensities range from VI-VII to VIII. The most important event occurred inside this area source is the M_W 6.8, March 31, 1969 Shedwan earthquake [16, 181]. Three foreshocks and 17 aftershocks (m_b 4.5-5.2) located in the Shedwan Island district are related to this big event. However, Maamoun and El-Khashab [182] mentioned that 35 foreshocks, taken place during the last half of March 1969, were preceding the main earthquake. The focal mechanism solutions of the largest two earthquakes (M_W 6.8, March 31, 1969 and M_W 5.5, June 28, 1972 earthquakes) show a normal faulting mechanisms with negligible shear component along the NW-trending fault plane that it is in agreement with the main axis of the Gulf of Suez [183]. This is also consistent with the results obtained using the waveform inversion techniques proposed by Huang and Solomon [184].

Central Gulf Of Suez (Eg-16) Seismogenic Source

The seismic activity in the EG-16 Central Gulf of Suez seismogenic source is relatively low when compared with the previous source. Five historical events are included in this source zone: the 1220 B.C., 1425, 1710, 1814, and 1879 earthquakes. Its intensities range from IV to VII. The most important earthquake inside this area was the M_S 6.2 March 6, 1900 event.

Abou Elenean [20] computed some focal mechanism solutions for earthquakes which taken place in the central part of the gulf, showing generally normal faulting, following the main gulf trend. A few of these events show slight strike-slip component, especially for those events closer to the transfer zones of the three gulf dip provinces [11]. This change, from a purely normal faulting in the southern part to a mixed (strike-slip and normal) movement, supports the separation between the southern and middle seismogenic zones in the Gulf of Suez.

Northern Gulf Of Suez (Eg-17) Seismogenic Source

Finally, the EG-17 Northern Gulf of Suez seismogenic source is characterized by its low seismic activity. Two large earthquakes occurred before the year 1900. They are the I_{max} VI, 742, and I_{max} V, 1754 earthquakes. Focal mechanism analyses for this seismogenic zone indicate normal faulting mechanism. Fault plane solutions by Abou Elenean [20] showed that the events located at the gulf apex show normal faults, generally trending NW-SE to WNW-ESE, and reflect a good agreement with the surface faults crossing the Eastern Desert from the gulf apex towards Cairo.

Abou Elenean [20] concluded that the focal mechanisms of small to moderate size earthquakes based on the P-wave polarities by Badway and Horváth [185-187], Badawy [188] and Salamon et al. [112], show the existence of few thrust faulting mechanisms along the Gulf of Suez trend. The author argues that these unexpected mechanisms could be due to the lack of local stations with clear polarities at that time. On the other hand, borehole breakouts analyses performed by Badawy [188] show a different stress direction, inconsistent with the NE-SW tension direction estimated from earthquake focal mechanisms.

Seismicity parameters for the delineated seismic sources along the Gulf of Suez are displayed in Table 3.

Table 3: b-value, annual rate of earthquakes, and maximum observed magnitude for the delineated seismic source zones along the Gulf of Suez.

Source Zone	b-value	Yearly Number of Earthquakes		Observed M_{max}
		Above $M_W 4.0$	Above $M_W 5.0$	
EG-15	1.06	0.8347	0.0721	M_W 6.8 on 1969/03/31
EG-16	0.8	0.3085	0.0488	M_S 6.2 on 1900/03/06
EG-17	0.86	0.1381	0.019	M_S 6.6 on 1754/--/--

Seismic Sources Of The Egyptian Eastern Desert

The Eastern Desert of Egypt, structurally, is a part of the Arabian-Nubian Shield. It lies within the fold and thrust belt of the Pan-African continental margin [189]. It is underlain mainly by the Pre-Cambrian basement of igneous and metamorphic rocks, which constitutes the Nubian Shield that had been formed before the Red Sea opening. It is believed that Nubian Shield basement was stabilized during the Pan-African Orogeny (about 570 Ma ago) [190].

Stern and Hedge [191] divided the Eastern Desert belt into three structural domains (Figure 2): northern, central and southern. These domains are separated by two major faults: i) the first is the Safaga-Qena zone, extending from Safaga to Qena, and ii) the second one is the Marsa Alam-Aswan fault zone. The Eastern Desert is characterized by E-W trending faults in the southern part, which changes to ENE-WSW in the middle one, near to Hurghada city. Further to the north, towards the Cairo-Suez District, the main fault trend becomes in the E-W direction.

However, Youssef [10] classified the main tectonic structures developed in the Eastern Desert into three main groups: i) NW-SE trending normal faults parallel to the Gulf of Suez-Red Sea Rift, ii) NE-SW trending faults parallel to the Gulf of Aqaba, and iii) a set of fault system trending nearly in the E-W direction. In addition, there are many simple and open folds with a NW-SE trend and low plunges.

Deif *et al.* [57] quote that the relationship between the earthquake activity in the Eastern Desert and the causal structures is not fully understood, due to the lack of geological and geophysical studies in this region. Furthermore, no historical earthquakes have been reported in the current seismogenic sources [17, 21]. The following seismogenic sources are identified (Figure 9).

Western Red Sea Coast (Eg-18) Seismicity Source
In addition to the Red Sea seismogenic sources mentioned above, there are some earthquakes located in the region which extends to the west, from the EG-12 Southern Egyptian Red Sea source till the western coast of the Red Sea. This activity may be related to the block adjustment in this region or to some ocean floor spreading. This source is characterized by a low seismic activity. The biggest observed earthquake is the M_L 4.5, May 23, 1990 earthquake.

Southern Eastern Desert (Eg-19) Seismicity Source
This seismicity source exhibit a low seismic activity rate in comparison to the adjacent Red Sea seismic sources. There are no focal mechanism solutions for earthquake events inside this area source. The M_L4.4, July 15, 1991 earthquake is the biggest recorded event in this zone.

Southern Abu Dabbab (Eg-20) Seismicity Source
Depending on both the changes in the seismicity rate and distribution, another seismicity source (EG-20) has been considered to the north of the previous zone. The same as the previous, there is no focal mechanism solutions in this source. The biggest recorded event is the M_L 4.7, January 21, 1982 earthquake.

Abu Dabbab (Eg-21) Seismicity Source
The Abu Dabbab region is located in the central part of the Eastern Desert of Egypt. The moderate level of seismic activity and extremely tight clustering of low-magnitude earthquakes at Abu Dabbab suggests that the seismicity in this area is not directly related to regional tectonics. One possible explanation is that the activity is related to magmatic intrusions into the Pre-Cambrian crust, but there is no direct evidence to support this hypothesis [161].

The most important event included inside this area source is the M_S 5.3, November 12, 1955 earthquake. This event is felt in the Upper Egypt in Aswan and Qena cities, and as far as Cairo, but no damage was reported. Its focal mechanism solution has normal and strike-slip faulting components produced by a NNW minimum compressive stress and a NE maximum compressive stress. Fault planes strike roughly E-W or N-S to NE-SW. Another important event related to this area is the M_W 5.1, July 2, 1984 earthquake, which is felt strongly in Aswan, Qena and Quseir cities. A large number of foreshocks and a huge sequence of aftershocks are recorded. The focal depth of the whole sequence was less than 12 km.

Seismicity parameters for the delineated seismic sources of the Eastern Desert of Egypt are displayed in Table 4.

Table 4: b-value, annual rate of earthquakes, and maximum observed magnitude for the delineated seismic source zones of the Eastern Desert.

| Source Zone | b-value | Yearly Number of Earthquakes | | Observed M_{max} |
		Above M_W 4.0	Above M_W5.0	
EG-18	1.29	0.1566	0.008	M_L 4.5 on 1990/05/23
EG-19	1.15	0.1182	0.0085	M_L 3.9 on 1991/07/15
EG-20	1.2	0.0625	0.0039	M_L 4.7 on 1982/01/21
EG-21	0.87	0.3714	0.0496	m_b 6.2 on 1955/11/12

Seismic Sources Along Nasser's Lake, Nile Valley And Cairo-Suez Region

Southern Aswan (Eg-22) Seismogenic Source
The geological structural pattern of the Nasser's Lake and Aswan region is characterized by a regional basement rock uplift and regional faulting

[192-197]. Faults around the Aswan region, according to their behavior, are grouped into three categories [82]:

i.   E-W trending faults (Figure 11), as the Kalabsha and Seiyal Faults, which lay to the west of Nasser's Lake. The Kalabsha Fault is about 185 km long right-lateral strike-slip fault. Its slip rate was estimated to be 0.028 mm/yr., and the Seiyal Fault is considered to be similar to that of the Kalabsha Fault [82].

ii.   N-S trending faults (Figure 11), which can be subdivided into two main sets: *The first set* lies to the NW of Nasser's Lake and consists of three faults: the Gebel El-Barqa Fault, the Kurkur Fault and the Khur El-Ramla Fault. The Gebel El-Barqa is a left-lateral strike-slip fault, with a total length of 110 km. The Kurkur Fault is also a left-lateral strike-slip fault, and it is characterized by its low seismic activity if compared with the neighbor faults. The Khur El-Ramla Fault is about 36 km in length, and it has no direct indication of its sense of movement. *The second set* of faults are lying to the SW of Nasser's Lake, and consists mainly of two faults: the Abu Dirwa and the Ghazala Faults. Abu Dirwa Fault is a 20 km long left-lateral strike-slip fault and it has a very low degree of seismic activity. In addition, for the Gazelle Fault, the analysis of its geomorphic expression shows no active features, and that there is no ground cracks observed along the fault trace. Likely this fault is inactive [192]. The fault planes of this system are nearly vertical (80-85°).

iii. The third one is a fault system trending NNE-SSW (Figure 11) and lies to the east of Nasser's Lake. The Dabud Fault, which represents the main fault of this group, is about 36 km length. Geological evidences indicate reverse-slip, opposed to the tectonic setting of the area.

In addition to the previous fault systems, Deif [57] has mentioned three faults located at the High Dam area. They are the Powerhouse, the Spillway and the Channel Faults. Deif [57] provided that the evidence of the occurrence of these faults is hidden below the Aswan High Dam and Nasser's Lake. These three faults show no evidence of being active in the Quaternary, and are considered as inactive with no significant hazard to the Aswan region [82].

No historical earthquakes were reported by Ambraseys *et al.* [18] inside this area source. However, two historical events (epicentral intensity VII) were reported by Maamoun *et al.* [16] to be located at the same place of the M_W 5.8, November 14, 1981 earthquake. These two events occurred in 1210 B.C. and in 1854.

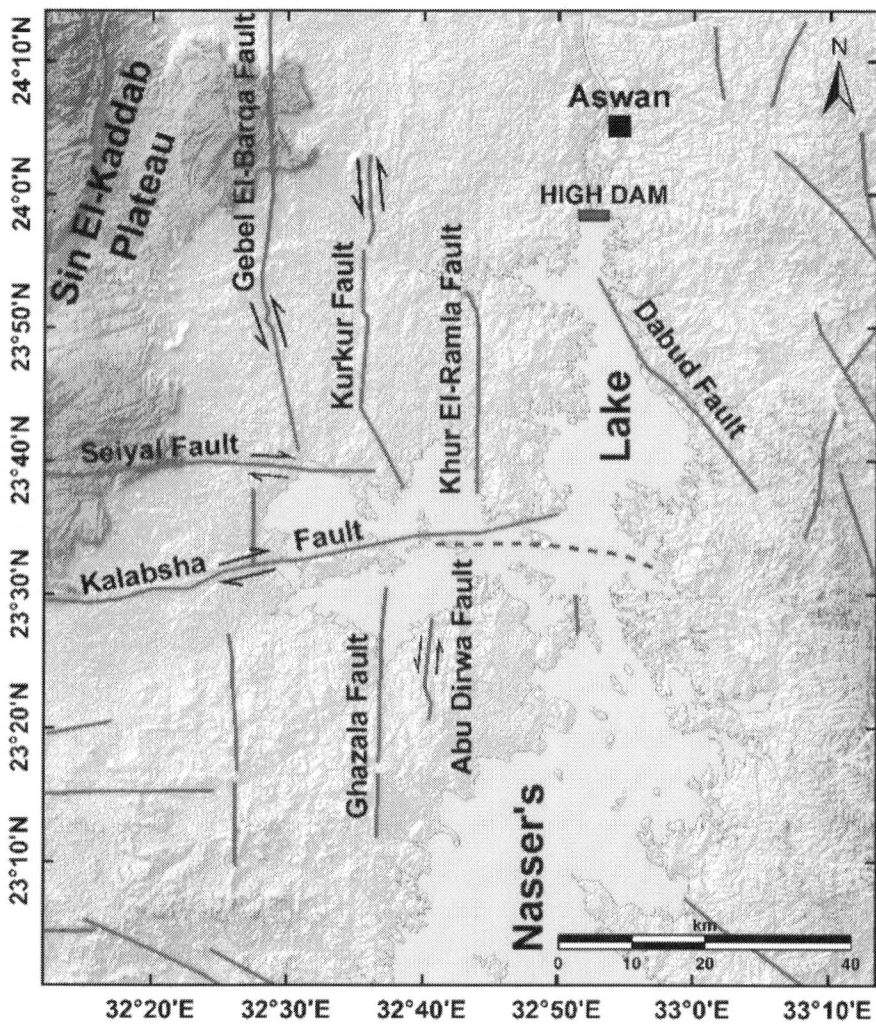

Figure 11: Geological and tectonic features around Nasser's Lake (redrawn after Woodward-Clyde Consultants [82]).

Woodward-Clyde Consultants [82] evaluated the fault system in the Kalabsha area and reported that the Western Desert Fault System consists of a set of E-W faults that exhibit dextral-slip displacement, and a set of N-S faults that exhibit sinistral-slip displacement. The E-W faults are longer, and have greater degree of activity in the Quaternary, having larger total slip rates (about 0.03 mm/yr.) than the N-S faults (0.01–0.02 mm/yr.).

Many seismic hazard studies have been carried out in the Aswan area and its surroundings due to its importance and neighborhood to the High

Dam (e.g., [51, 57 and 58]). Three alternative seismotectonic models for Aswan area have been considered in these studies. The first model consider the Aswan Area as one seismotectonic model, while in the second one is subdivided into six seismotectonic provinces. The third model is mainly depending on the fault seismic sources. The latter one is based mainly on the well-known defined active faults and its associated seismic activities.

However, this work, the Aswan region and its surroundings is considered as one source zone (Figure 9). The main earthquake that took place inside this area was the M_W 5.8, November 14, 1981 event. This earthquake occurred in the Nubian Desert of Aswan. It is of great significance because of its possible association with Nasser's Lake. Its effects were strongly felt up to Assiut city (440 km to the north from Kalabsha Fault), as well as to Khartoum city (870 km to the south). Several cracks on the western bank of the Nasser's Lake, and several rock-falls and minor cracks on the eastern bank, are reported. The largest of these cracks is about one meter in width and 20 km in length. This earthquake was preceded by three foreshocks and followed by a large number of aftershocks. The focal depth of this earthquake is estimated to be 25 km. The composite fault plane solution of this event indicates a nearly pure strike-slip faulting with a normal-fault component [49, 195].

Luxor- Southern Beni Suef (Eg-23) Seismogenic Source

Several geophysical studies have been carried out by many authors using different approaches in individual localities lying along the Nile Valley. The most interesting geological studies in the Nile Valley are those carried out by Said [69-71], Issawi [78], Philobbos *et al.* [86], and El-Younsy *et al.*[196]. All these works were conducted independently and aimed to obtain information about the drainage system, the stratigraphy and structural geology in this part of Egypt.

The Nile Valley is a large elongated Oligo-Miocene rift, trending N-S as an echo of the Red Sea rifting. There is no agreement among scientists, till now, about the origin of the Nile Valley. Some authors [197, 198] supported the opinion of the erosional origin of the Nile Valley, while many others (e.g., [11, 12, 13, 17, 70 and 199]) consider the tectonic origin. This is supported by the fault scarps bordering the cliffs of the Nile Valley, the numerous faults recognized on its sides [70, 71 and 199] and the most recent focal mechanism solutions. Furthermore, geological studies of the Nile Valley show that, it occupies the marginal area between two main tectonic blocks (the Eastern Desert and the Neogene-Quaternary platform), which in turn behaves as a barrier that prevents the further extension of the East African Orogenic Belt activity to the west [71].

From the structural point of view, the faults and joints are the most deformational features observed at the cliffs bordering the Nile stream [69 and 70]. These faults have different directions (Figure 2). The most abundant present the NW-SE and NNW-SSE trends, while others (less abundant) exhibit the WNW-ESE, ENE-WSW and NE-SW directions. Most of the major valleys, at the east of the Nile River, are generated and controlled in a more or less degree by these faults.

To the north of Aswan area, in the region between Luxor and Southern Beni Suef, along the Nile River, there is a low seismicity level, which coincides with the main trend of the Nile River. This active area has been considered as a separate seismogenic source. Several historical earthquakes are reported to occur along the Nile River in this area source that may be due to the high population density along the Nile River in the ancient times. These earthquakes are the 600 B.C., 27 B.C., 857, 967, 997, 1264, 1299, 1694, 1778, and 1850 events. Their intensities range from V to VIII. Focal mechanism solutions exhibit reverse faulting mechanism to the west of the Nile River, in the area between Luxor and Assiut. However, normal faulting mechanism with strike-slip component appears to the north of Assiut till Beni Suef city.

Beni Suef – Cairo – Suez District (Eg-24) Seismogenic Source

To the north of the previous zone and to the west of the Gulf of Suez, there is a moderate seismic activity between Beni Suef and Cairo, on the River Nile, till Suez, on the apex of the Gulf of Suez (Figure 9). Three fault trends are affecting the Cairo-Suez district: the first one is trending E-W, which aligned by latitude 30°N, and it is very dominant, while the other two (ENE and NW) are spatially more abundant [200]. The faults are predominantly normal, and have produced a series of fault blocks with a large strike-slip component [200].

Field observations, satellite images, aerial photographs and seismic profiles confirm that the region between Cairo and Suez is active from a tectonic point of view. Seismic activity are noticed along this belt at Wadi Hagul and Abu Hammad. However, the earthquake distribution in this area is very scattered, and cannot be attributed to a specific known fault. This disperse seismicity yields a difficulty in delineating seismic zones. It is assumed that the seismic potential is uniform throughout the zone, although this is not entirely clear.

Sixty one historical earthquakes are related to this area source. The most important among them are the 935, 1111, 1259, 1262, 1303, and 1588 events. Moreover, the most important instrumental earthquake taken place in this source is the M_W 5.8, October 12, 1992 event. Its epicenter was located about 40 km south of Cairo, in Dahshour. It caused a disproportional damage (estimated at more than L.E. 500 million) and the loss of many lives. The shock was strongly felt, and caused sporadic damage and life loss in the Nile Delta, around Zagazig. Damage was extended to reach Fayoum, Beni Suef and Minia cities. The mostly affected area was Cairo, especially its old sections, Bulaq and the southern region, along the western bank of the Nile to Gerza (Jirza) and El-Rouda. In all, 350 buildings collapsed completely and 9000 were irreparably damaged, killing 545 persons and injuring 6512. Most causalities in Cairo were victims of the horrible stampedes of students rushing out from schools. Approximately, 350 schools and 216 mosques were destroyed and there was about 50000 homeless.

Abdel Tawab *et al.* [201] studied the surface tectonic features of the area around Dahshour and Kom El-Hawa, and found a major N55°E trending normal fault at Kom El-Hawa (800 m length of surface trace with a vertical displacement of 40 cm) and a major E-W trending open fracture at Dahshour area (1200 m in length). Maamoun *et al.* [202] concluded that, most of the surface lineaments recorded after the occurrence of the main shocks are trending E-W to NW-SE. Abou Elenean [19] studied the focal mechanism solutions for some earthquakes in Dahshour area, and found normal faulting with a large strike-slip component. The first nodal plane is trending nearly E-W, showing coincidence with the surface lineaments that appeared directly after the occurrence of the M_S 5.9, 1992 earthquake.

In addition to the previous earthquake, there were three important earthquakes located inside this source zone. One of them located to the southwest of Suez, is the m_b 4.9 (M_S 4.8), March 29, 1984 Wadi Hagul earthquake. It was strongly felt in Suez, Ismailia and Cairo. A large number of aftershocks were recorded by nearby temporary stations. The second earthquake was located Northeast Cairo; it is the m_b 4.8, April 29, 1974 Abu Hammad earthquake. It was strongly felt in Lower Egypt (Nile Delta) and Southwest Israel. The last earthquake was the m_b 5.0, January 2, 1987 Ismailia event.

Mousa [203] and Hassib [204] computed two nodal planes trending ENE-WSW and NNW-SSE, with left-lateral strike-slip motion along the second plane, for the Abu Hammad event. They computed the same strike-slip with reverse component for the Wadi-Hagul earthquake. In addition, the mechanism of the Ismailia earthquake shows also strike-slip movement with two nodal planes trending N68° E and S24°E, with steep dip angles (each of them is 80°) [205].

The seismicity parameters for the delineated seismic sources along the Nile River are displayed inTable 5.

Table 5: b-value, annual rate of earthquakes, and maximum observed magnitude for the delineated seismic source zones along the Nile River.

| Source Zone | b-value | Yearly Number of Earthquakes | | Observed M_{max} |
		Above M_W 4.0	Above M_W5.0	
EG-22	0.79	0.586	0.0946	M_W 5.8 on 1981/11/14
EG-23	0.73	0.2948	0.0549	I_{max} VIII on 0857/04/--
EG-24	0.99	0.5964	0.0608	I_{max} IX-X on 1262/--/--

Seismic Sources Along The Mediterranean Coastal Line

The Mediterranean Coastal area is characterized by small to moderate seismicity. This area is located at the southeastern part of the Mediterranean Sea. It separates between the high seismic activities along the Gulf of Aqaba-Dead Sea Transform Fault and the seismicity of the Mediterranean Sea (Hellenic and Cyprian Arcs). Moreover, it separates the Southern Cyprus seismic activity from the Northern Egypt activity. Hence, this area has been divided into three seismic sources (Figure 12), based mainly on the available focal mechanism data and the seismic activity.

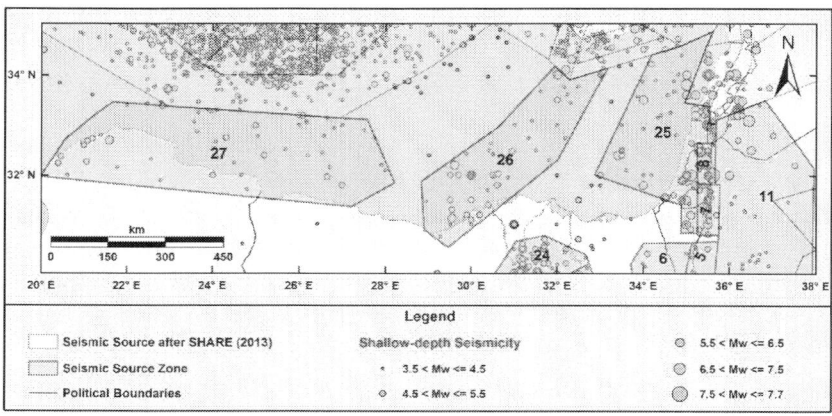

Figure 12: Shallow-depth seismicity (h ≤ 35 km) and seismic source zones delineated for the Mediterranean coastal line.

Eastern Mediterranean Coast (Eg-25) Seismicity Source

This area source is parallel to the eastern coastal line of the Mediterranean Sea. It is located to the west of the previous quoted sources EG-07, EG-08, and EG-09, and to the southeast of Cyprus. It includes all the seismicity located to the west of the DST, and those earthquakes are not related with the Cyprian Arc. 29 historical events are included inside this area source. The most important among them are the 590 B.C., 525 B.C., 12 B.C., 306, 332, 551, 1269, and 1546 earthquakes.

Northern Delta (Eg-26) Seismicity Source

This source is located to the northwest of the Nile Delta region. It extends from Alexandria towards the Mediterranean Sea in NE direction. 23 historical events are included inside this large area source, among them the 796, 951, 955, 956, 1303, 1341, and 1375 earthquakes. Moreover, the m_b 6.5 (M_S6.8), September 12, 1955 Alexandria earthquake, represents the most important recorded event inside this source. This earthquake was felt in the entire Eastern Mediterranean Basin. In Egypt, it was strongly felt, and led to the loss of 22 lives and damage in the Nile Delta, between Alexandria and Cairo [17]. The destruction of more than 300 buildings of old brick construction was reported in Rosetta, Idku, Damanhour, Mohmoudya and Abu-Hommos. A maximum intensity of VII was assigned to a limited area in Behira province, where 5 people killed and 41 were injured.

Mostly of the focal mechanism data inside this area source reflects reverse faulting mechanism with, sometimes, strike-slip component, except one event, showing a strike-slip motion with a notable normal component (the M_W 4.5, April 9, 1987 event).

Western Mediterranean Coast (Eg-27) Seismicity Source

The Western Egyptian Mediterranean Coastal zone is located to the north of the Egyptian-Libyan boundary. Only two historical events are reported inside this source: the I_{max} VIII, 262, and I_{max} VI, 1537 earthquakes. However, the most important recorded earthquake is the M_W 5.5, May 28, 1998 Ras El-Hekma event. This earthquake is widely felt in Northern Egypt. Intensity of VII is assigned at Ras El- Hekma village (~300 km west of Alexandria), and an intensity of V–VI at Alexandria city [206]. Ground fissures trending NW–SE were observed along the beach. Some cracks were also observed in concrete buildings. Furthermore, some people left their houses. The windows rattled and hanging objects swung, but the direction of the ground motion was poorly identified [206].

Recent studies concerning the crustal structure and focal mechanism of the M_W 5.5, May 28, 1998 Ras El-Hekma earthquake suggested that this source is an extension of the compressional stress from the Hellenic Arc. This compressional stress reactivated the old Triassic normal faults as reverse faults, or reverse faults with strike-slip component. This activity coincides with the hinge zone geometry proposed by Kebeasy [17]. Mostly of the focal mechanism analyses data indicate reverse faulting with some strike-slip component.

Seismicity parameters for the delineated seismic sources along the Mediterranean coastal line are displayed in Table 6.

Table 6: b-value, annual rate of earthquakes, and maximum observed magnitude for the delineated seismic source zones along the Mediterranean coastal line.

Source Zone	b-value	Yearly Number of Earthquakes Above M_W 4.0	Above M_W5.0	Observed M_{max}
EG-25	0.97	0.5204	0.0554	I_{max} X on 1546/01/14
EG-26	0.94	0.5975	0.0688	m_b 6.5 on 1955/09/12
EG-27	0.6	0.7134	0.1793	I_{max} VIII on 0262/--/--

Seismic Sources Of The Western Desert

El-Gilf El-Kebeir (Eg-28) Seismogenic Source

Issawi [28] studied the geology of El-Gilf El-Kebeir region, and concluded that the area is affected by three main faults (Figure 2). The first one is the Gilf Fault, which strikes N-S for a distance of 150 km inside Egypt. Its extension in Sudan is unknown. Its northward extension is not traced. He interpreted this fault as a normal, gravity, strike and hinge type of structure. The second one is Kemal Fault, which limits the northwestern side of the Gilf Plateau. It is normal, strike fault which trends NW-SE. The Kemal Fault intersects the Gilf Fault at its northern end. The third one is the Tarfawi Fault, which has the same trend similar to the Gilf Fault. Its length, in Egypt, is 220 km but it extends in Sudan. He interpreted this fault as a normal, gravity and hinge fault.

The only recorded earthquake in this area source is the m_b 5.3 (M_L 5.7), December 9, 1978 El-Gilf El-Kebeir earthquake. It had a reverse faulting mechanism. Riad and Hosney [207] studied its focal mechanism and concluded that, a shear direction did exist in the basement rocks of the southern part of the Western Desert and has been explained as due to compressional stress resulting from the spreading of the Red Sea. Their fault planes solution shows that the P-axis is almost perpendicular to the

Red Sea spreading axis. They concluded that the Gilf Plateau is probably divided into two parts by a fault striking nearly E-W. Some authors [e.g., 208 and 209] pointed out that this activity is linked to the pre-existing weak zones, while, Abou Elenean [19] linked such an intraplate activity to the intersection of more than one local fault.

In the current work, the Gilf El-Kebeir (EG-28) seismogenic source covers the seismic activity in this area, as well as the above mentioned faults.

EASTERN MEDITERRANEAN REGION SEISMIC SOURCES

The Mediterranean region is characterized by a very complex tectonics that can be generally described in the frame of the collision between the Eurasian and African Plates [183, 210-219]. It can be divided into western, central, and eastern basins.

The Eastern Mediterranean region, which defines the region lying between the Caspian Sea and the Adriatic Sea through Caucasus, Anatolia, Aegean Sea and Greece, is one of the world's most seismically active regions. Recent tectonics of the Eastern Mediterranean region has been studied intensely in the last four decades. The Eastern Mediterranean region is known to be seismically active over a period of more than 2000 years based on historical and instrumental records. The tectonic and seismotectonic studies reflect a highly complicated tectonic setting.

It is characterized by two main seismic regions: the Hellenic and Cyprian Arcs (Figure 13). The Cyprian Arc has a similar geometry to the Hellenic Arc and the two are often compared (e.g., [220]). However, the observed seismic activity and the well-known plate movement in the Eastern Mediterranean area, suggest that the previously mentioned arcs are affected by a very distinct tectonic activity. The convergence across the first one (Hellenic Arc) is 20–40 mm/yr. (two to three times faster across the Cyprian Arc). Thus, this biggest displacement level yields in higher seismicity rate at much deeper levels (up to 300 km) [220].

The Cyprian Arc represents a tectonic plate margin separating the Anatolian sub-Plate (to the north) from the Nubian and Sinai sub-Plates (to the south) (Figure 13). It is connected from the west by the Hellenic Arc, and from the east by the Dead Sea and the East Anatolian Faults. In addition, it extends from the Gulf of Antalia, to the west to the Gulf of Iskenderun, to the east. On the other hand, the Hellenic Arc is considered to be the most active seismic region in Europe. It is represents the convergent plate boundary between the African Plate and the Eurasian Plate (Aegean sub-Plate) in the Mediterranean area (Figure 13).

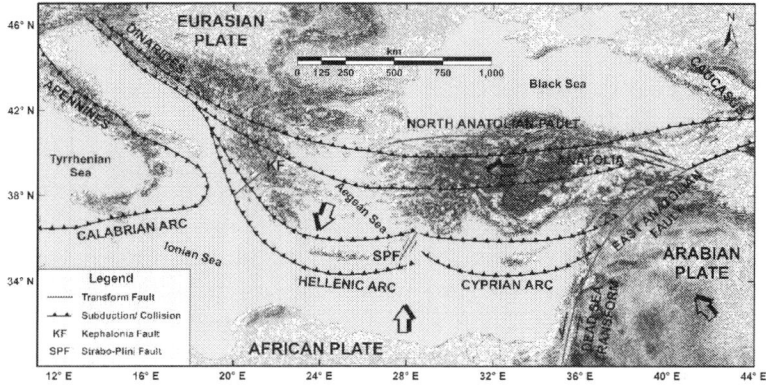

Figure 13: Tectonic map of the Eastern Mediterranean region (after Ziegler [221]; Meulenkamp *et al.* [222]; and Dewey *et al.* [223]).

Share Shallow-Depth Seismic Sources (H ≤ 20 Km)

The Seismic Hazard Harmonization in Europe (SHARE) project [224], since the year 2009 till 2013, worked in establishing an appropriate seismic hazard model for Europe and Turkey. This project delivered a seismic hazard reference model for the current use of the European building design and seismic regulations, Eurocode 8 (EC8), that came into force in 2010.

The EU-FP7 European Commission Project (SHARE), aiming at providing an updated state-of-the art time-independent seismic hazard model, envisioned to serve as a reference model for the revision of the EC8 building code. SHARE, in addition, contributes its results to the Global Earthquake Model (GEM, www.globalquakemodel.org), a public/private partnership initiated and approved by the Global Science Forum of the OECD- GSF, aiming to provide a uniform hazard and risk model around the globe.

The Euro-Mediterranean area is complex from a seismotectonic point of view. The plate boundary between Africa and Europe runs roughly west to east from the Mid-Atlantic Ridge to Eastern Turkey with different mechanisms including continental collision, subduction, and transcurrent movement. Moving away from the plate boundary, the stable continental region is also locally rather active.

SHARE inherits knowledge from national, regional and site-specific PSHAs, assessed new data, assembled the data in a homogeneous fashion, and built comprehensive hazard relevant databases. In the frame of this project, the establishment of a seismic source model for Europe and the surrounding areas was considered. This model is built upon the available local and regional models as well as newly defined source zones. It has been developed during eight separate workshops by the SHARE consortium. Almost 80 experts from 28 countries from the informed European-Mediterranean seismological community have participated in building the zonation model.

The principle for seismic source zones is that they represent enclosed areas within which, a uniform seismicity distribution and maximum magnitude is expected. Background sources have been avoided in the sense that all areas have been covered by seismic sources, i.e., even very low seismicity areas are covered with areal source zones. The principles along which seismic source zones in the current model have been constructed are based on information from geological structures on different scales, tectonics and seismicity.

Seismicity also follows these structures well, e.g., as can be seen along the North Anatolian Fault, the Gulf of Corinth and the Hellenic Arc. The use of fault source information has also been done in the delineation of the source zones, especially in the case of the foundation of the sources for Balkans, Greece and Turkey, Italy and Portugal. b-value, annual activity rates, and maximum expected magnitude were computed using different approaches and methods and included in the SHARE project database (www.share-eu.org) [225].

In the current work, 53 shallow-depth seismic sources (h ≤ 20 km) from the SHARE source model (Figure 5), were considered to the north of Egypt, till latitude 38° N, and covering the Greece and Turkey regions. Some of the events located at this region were felt and caused few damages in the northern part of Egypt (e.g., the I_{max} VIII, August 8, 1303 offshore Mediterranean earthquake, the I_{max} VI, February 13, 1756 and the M_S 7.4, June 26, 1926 Hellenic Arc earthquakes, the M_W 6.8, October 9, 1996 Cyprus earthquake, and the M_S 6.4, October 12, 2013 Crete earthquake). Thus, these source zones have a certain contribution to the seismic hazard in Northern Egypt.

The model in the Greek and Cyprian area build to a large extent upon the previous works of Papiouannou and Papazachos [64] and Papiouannou [226]. The Turkish model [227, 228] is provided as a cooperation between the EMME project and SHARE. The Turkish model is further a hybrid model, in the sense that the area sources have been delineated with respect to the integrated fault line sources from the main faults, like the North and East Anatolian Faults.

Intermediate-Depth Seismic Sources (20 ≤ H ≤ 100 Km)

Intermediate focal depth earthquakes occur in the Eastern Mediterranean region (Southern Greece and Turkey) define a Benioff zone of stair shape which dips from the convex side of the Cyprian and the Hellenic Arcs to its concave side (from the Eastern Mediterranean to the Greek and Turkish lands) [229-231]. Some of these earthquakes are moderate to large earthquakes, and constitute a seismic threat for the whole Mediterranean area, including Northern Egypt. Since, because of their magnitudes and focal depths, these earthquakes produce seismic waves of large amplitude and period which travel large distances with low attenuation [232]. Therefore, these earthquakes can contribute to the seismic hazard of Northern Egypt.

In this work, intermediate-depth sources for earthquakes having focal depths ranging from 20 km to 100 km have been delineated. Below this depth (100 km) and considering the large distance from Egypt, deep events have no contribution to the seismic hazard. Thus, 7 intermediate-depth source zones have been considered in the Hellenic and Cyprian subduction zones to cover the intermediate-depth seismicity ($20 \leq h \leq 100$ km) (Figure 14). The zoning was based on the seismicity distribution and the tectonic setting of the region. Seismicity parameters for these intermediate-depth seismic sources are displayed in Table 7.

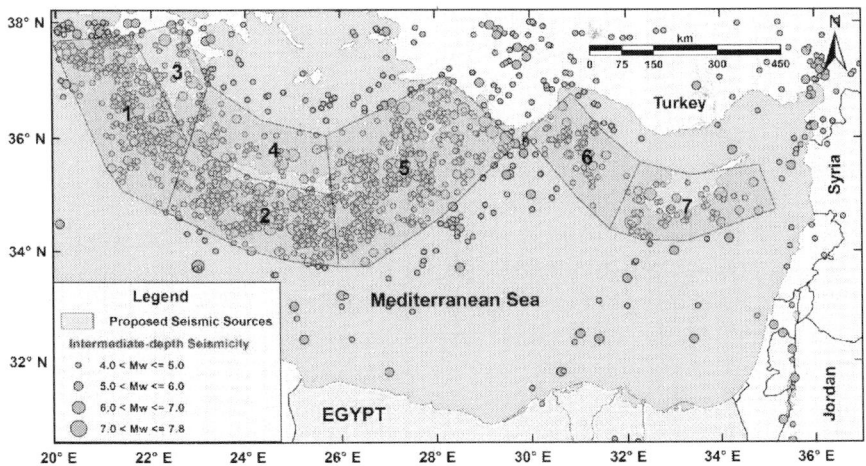

Figure 14: Intermediate-depth seismicity ($20 \leq h \leq 100$ km) and delineated intermediate-depth seismogenic source zones for the Eastern Mediterranean region.

Table 7: b-value, annual rate of earthquakes, and maximum observed magnitude for the delineated seismic source zones of the Eastern Mediterranean region.

Source Zone	b-value	Yearly Number of Earthquakes Above M_W 4.0	Above M_W 5.0	Observed M_{max}
MD-01	0.88	12.2091	1.6212	M_W 6.9 on 2008/02/14
MD-02	0.97	14.1395	1.5283	m_b 6.4 on 2013/10/12
MD-03	0.78	1.9892	0.3324	m_b 6.9 on 1962/08/28
MD-04	0.83	2.8908	0.4316	m_b 7.2 on 1903/08/11
MD-05	0.88	14.8782	1.9434	m_b 7.1 on 1984/02/09
MD-06	0.93	2.2986	0.2727	M_S 5.8 on 1984/04/30
MD-07	0.82	2.9083	0.4409	m_b 6.5 on 1941/01/20

CONCLUSION

To reach a more realistic seismic hazard quantification in Egypt, it is necessary to recognize the seismic source zones, including the seismic activity that can affect different regions all over the country. In the current work, a new seismic source model for Egypt and its surroundings is proposed, using all available geological, geophysical, tectonic and earthquake data, aimed at carrying out seismic hazard studies.

This work presents a detailed review on major tectonic features and the correlation of seismicity with them, to demarcate seismic sources in Egypt and neighborhood. The Gulf of Aqaba-Dead Sea Transform, the Red Sea-Gulf of Suez Rift, and the Cyprian and Hellenic Arcs are the three most active seismotectonic belts in the region, which have produced several large earthquakes in the recent past. On the basis of a comprehensive and critical analysis of the seismotectonic characteristics, different seismic sources are defined to model the seismicity for the assessment of seismic hazard in Egypt.

Focal mechanism solutions data, active faults data, as well as an updated earthquake catalogue for the period 2200 B.C.–2013 are taken into account. Potential seismic sources are modeled as area sources, in which configuration of each seismic source is controlled, mainly, by the fault extension and seismicity distribution.

The proposed seismic source model consists of 28 shallow-depth seismic zones (h ≤ 35 km) for the Egyptian territory and its surroundings, specified on the basis of mainly seismotectonic and seismicity criteria. In addition, the authors have considered 53 shallow-depth seismic sources (h ≤ 20 km) for the Eastern Mediterranean region after SHARE (2013). Furthermore, the current model involves 7 delineated intermediate-depth seismic sources (20 ≤ h ≤100 km) covering the intermediate-depth seismicity in the Eastern Mediterranean region.

Seismicity parameters (b-value and activity rates) of the Gutenberg–Richter magnitude–frequency relationship have been estimated for each one of the seismic sources. In addition, the maximum observed magnitude for each seismic source zone was reported from the sources sub-catalogues. The coordinates of these seismic source zones and the estimated seismicity parameters can be directly inserted into PSHA after the estimation of the maximum expected magnitude for each source. The computation of seismic hazard for Egypt using these data will form the subject matter of a future paper.

ACKNOWLEDGEMENTS

The first author wants to thank the Egyptian Government for funding him in the Joint-Supervision Mission program at the University of Jaén, Spain. This research was supported by the Aswan Regional Earthquake Research Centre and the Spanish Seismic Hazard and Active Tectonics research group.

REFERENCES

1. Cornell CA. Engineering Seismic Risk Analysis. Bulletin of Seismological Society of America 1968; 18: 1583-1606.
2. Reiter L. Earthquake Hazard Analysis. Columbia University Press, 1990.
3. McGuire RK. FORTRAN Computer Programs for Seismic Risk Analysis. United States Geological Survey. Open-File Report No. 76-67, 1976.
4. Kramer SL. Geotechnical Earthquake Engineering. Prentice-Hall Editor. Upper Saddle River, New Jersey, 07458; 1996.

5. Abrahamson N. Seismic Hazard Assessment: Problems with Current Practice and Future Development. First European Conference on Earthquake Engineering and Seismology, 3-8 September 2006, Geneva, Switzerland.

6. McGuire RK. Computer Program for Seismic Risk Analysis Using Faults as Earthquake Sources (FRISK) 1978. US. Department of Interior Geological Survey. Open-File Report 78-1007, 71 pp., Denever, Colorado.

7. Bender BK. and Perkins DM. SEISRISK III: A Computer Program for Seismic Hazard Estimation. USGS Bulletin 1987: 1772.

8. Ordaz M., Faccioli E., Martinelli F., Aguilar A., Arboleda J., Meletti C., and D'Amico V. CRISIS 2014. Institute of Engineering, UNAM, Mexico City, Mexico.

9. Woo G. NPRISK Seismic Hazard Computation Algorithm Based on Cornell-McGuire Principles. Code developed at NORSAR; 1994.

10. Youssef MI. Structural Pattern of Egypt and its Interpretation. The American Association of Petrolum Geologists Bulletin 1968; 53: 601-614.

11. Meshref W. Tectonic Framework. In: Said R. (ed.). The Geology of Egypt. A.A. Balkema, Rotterdam, Netherlands; 1990. p113-155.

12. Sieberg A. Handbuch der Geophysik. Band IV, Erdbeben-geographie. Borntraeger, Berlin, 1932a. p527-1005.

13. Sieberg A. Erdbeben und Bruchschollenbau in Östlichen Mittelmeergebiet. Denkschriften der Medizinisch-Naturwissenschaftlichen Gesellschaft zu Jena 18, No. 2; 1932b.

14. Ismail A. Near and Local Earthquakes at Helwan from 1903-1950. Helwan Observatory Bulletin No. 49, 1960.

15. Gergawi A. and Khashab A. Seismicity of U.A.R. Helwan Observatory Bulletin No. 76, 1968.

16. Maamoun M., Allam A. and Megahed A. Seismicity of Egypt. Bulletin of Helwan Institute of Astronomy and Geophysics, 109-160, 1984.

17. Kebeasy RM. Seismicity. In: Said R. (ed.) The Geology of Egypt. A.A. Balkerma, Rotterdam, Netherlands; 1990. p51-59.

18. Ambraseys NN., Melville CP., and Adams RD. The seismicity of Egypt, Arabia and Red Sea. Cambridge University Press; 1994.

19. Abou Elenean K. Seismotectonics of Egypt in relation to the Mediterranean and Red Sea tectonics. PhD. thesis. Ain Shams University, Egypt; 1997.

20. Abou Elenean K. Focal Mechanism of Small and Moderate Size Earthquakes Recorded by the Egyptian National Seismic Network (ENSN), Egypt. NRIAG Journal of Geophysics 2007; 6 (1) 119-153.

21. Badawy A. Historical Seismicity of Egypt. Acta Geodaetica et Geophysica Hungarica 1999; 34 (1-2) 119-135.

22. Badawy A. Seismicity of Egypt. Seismological Research Letters 2005; 76 (2) 149-160.

23. Abou Elenean KM., Mohamed AME., and Hussein HM. Source Parameters and Ground Motion of the Suez-Cairo Shear Zone Earthquakes, Eastern Desert, Egypt. Natural Hazards 2010; 52: 431-451.

24. Ziegler MA. Late Permian to Holocene Paleofacies Evolution of the Arabian Plate and its Hydrocarbon Occurrences. GeoArabia 2001; 6 (3) 445-504.

25. Pollastro RM. Total Petroleum Systems of the Paleozoic and Jurassic, Greater Ghawar Uplift and Adjoining Provinces of Central Saudi Arabia and Northern Arabian-Persian Gulf. US. Geological Survey Bulletin 2202-H; 2003.

26. Egyptian Geological Survey and Mining Authority "EGSMA". Geologic Map of Egypt 1:2000000; 1981.

27. Riad S. Shaer Zones in North Egypt Interpreted from Gravity Data. Geophsyics 1977; 24 (6) 1207-1214.

28. Issawi B. New Findings on the Geology of Uweinat, Gilf Kebir, Western Desert, Egypt. Annals of the Geological Survey of Egypt 1978; 8: 275-293.

29. Maamoun M. and Ibrahim EM. Tectonic Activity in Egypt as Indicated by Earthquake. Helwan Institute of Astronomy and Geophysics. No. 170, 1978.

30. Albert RNH. Seismicity and Earthquake Hazard at the Proposed Site for a Nuclear Power Plant in the El-Dabaa Area, North Western Desert, Egypt. Acta Geophysica Polonica 1986; 34 (3) 263-281.

31. Albert RNH. Seismicity and Earthquake Hazard at the Proposed Site for a Nuclear Power Plant in the Anshas Area, Nile Delta, Egypt. Acta Geophysica Polonica 1987; 35 (4) 343-363.

32. Kebeasy RM., Maamoun M., and Albert RNH. Earthquake Activity and Earthquake Risk around the Alexandria Area in Egypt. Acta Geophysica Polonica 1981; 29 (1) 37-48.

33. Kebeasy R. Seismicity of Egypt. Personal Communication; 1984.

34. Marzouk I. Study of the Crustal Structure of Egypt Deduced from Deep Seismic and Gravity Data. PhD. thesis. Institute of Geophysics, University of Hamburg; 1988.

35. Fat-Helbary RE. Investigation and Assessment of Seismic Hazard in Egypt. Unpublished Report Submitted to MAPFRE, Spain; 1999.

36. Fat-Helbary RE. Assessment of Seismic Hazard and Risk in Aswan Area, Egypt. PhD. thesis. Tokyo University, Japan; 1994.

37. Fat-Helbary RE. Probablistic Analysis of Potential Ground Motion Levels at the Principal Cities in Upper Egypt. Journal of Applied Geophysics 2003; 2: 279-286.

38. Reborto P., Paolo L., and Dimitris D. Seismotectonic Regionalization of the Red Sea Area and its Application to Seismic Risk Analysis. Natural Hazard 1992; 5: 233-247.

39. Mohammed A. Seismic Microzoning Study and its Applications in Egypt. PhD. thesis. Ain Shams University, Egypt; 1993.

40. EL-Hadidy S. Crustal Structure and its Related Causative Tectonics in Northern Egypt using Geophysical Data. PhD. thesis. Ain Shams University, Egypt, 1995.

41. Fat-Helbary RE., and Ohta Y. Assessment of Seismic Hazard in Aswan Area, Egypt. 11th World Conference on Earthquake Engineering. Paper No. 136 Published by Elsevier Science Ltd 1996.

42. El-Sayed A. and Wahlström R. Distribution of the Energy Release, b-values and Seismic Hazard in Egypt. Natural Hazards1996; 13: 133-150.

43. Abou Elenean K. Seismotectonics Studies of El-Dabaa and its surroundings. Unpublished Report 2010. NRIAG, Egypt.

44. Badawy A. Earthquake Hazard Analysis in Northern Egypt. Acta Geodaetica et Geophysica Hungarica 1998; 33 (2-4) 341-357.

45. Deif A. Seismic Hazard Assessment in and around Egypt in Relation to Plate Tectonics. PhD. thesis. Ain Shams University, Egypt, 1998.

46. Riad S., Ghalib M., El-Difrawy MA., Gamal M. Probabilistic Seismic Hazard Assessment in Egypt. Annals of the Geological Survey of Egypt 2000; 23: 851-881.

47. Abou ELenean K. and Deif A. Seismic Zoning of Egypt. Unpublished Work 2001. NRIAG, Egypt.

48. El-Sayed A., Vaccari V. and Panza GF. Deterministic Seismic Hazard in Egypt. Geophysical Journal International 2001; 144: 555-567.

49. Fat-Helbary RE., and Tealeb AA. A Study of Seismicity and Earthqauake Hazard at the Proposed Kalabsha Dam Site, Aswan, Egypt. Natural Hazards 2002; 25: 117-133.

50. El-Amin EM. Study of Seismic Activity and its Hazard in Southern Egypt. MSc. thesis. Assiut University, Egypt; 2004

51. El-Amin EM. Study of Seismic Hazard Analysis Using Fault Parameter Solutions in Aswan Region, Upper Egypt. PhD. thesis. Assiut University, Egypt; 2011.

52. El-Hefnawy M., Deif A., El-Hemamy ST., and Gomaa NM. Probablistic Assessment of Earthquake Hazard in Sinai in Relation to the Seismicity in the Eastern Mediterranean Region. Bulletin of Engineering Geology and the Environment 2006; 65: 309-319.

53. Abdel-Rahman K., Al-Amri AMS. and Abdel-Moneim E. Seismicity of Sinai Peninsula, Egypt. Arabian Journal of Geosciences 2009; 2 (2) 103-118.

54. El-Hadidy M. Seismotectonics and Seismic Hazard Studies for Sinai Peninsula, Egypt. MSc. thesis. Ain Shams University, Egypt; 2008.

55. El-Hadidy M. Seismotectonics and Seismic Hazard Studies in and around Egypt. PhD. thesis. Ain Shams University, Egypt; 2012.

56. Deif A., Abou Elenean K., El-Hadidy M., Tealeb A. and Mohamed A. Probabilistic Seismic Hazard Maps for Sinai Peninsula, Egypt. Journal of Geophysics and Engineering 2009; 6: 288-297.

57. Deif A., Hamed H., Igrahim HA., Abou Elenean K., and El-Amin EM. Seismic Hazard Assessment in Aswan, Egypt. Journal of Geophysics and Engineering 2011; 8: 531-548.

58. Fat-Helbary RE., El Khashab HM., Dojcinovski D., El Faragawy KO., and Abdel-Motaal AM. Seismicity and Seismic Hazard Analysis in and around the Proposed Tushka New City Site, South Egypt. Acta Geodynamica et Geomaterialia 2008; 5 (4) 389-398.

59. Mohamed AA., El-Hadidy M., Deif A., and Abou Elenean K. Seismic Hazard Studies in Egypt. NRIAG Journal of Astronomy and Geophysics 2012; 1: 119-140.

60. Sawires R., Ibrahim HA., Fat-Helbary RE., and Peláez JA. A Seismological Database for Egypt Including Updated Seismic and Focal Mechanism Catalogues. 8th Spanish-Portuguese Assembly of Geodesy and Geophysics. 29-31 January 2014, Évora, Portugal.

61. Papazachos BC. Seismicity of the Aegean and Surrounding Area. Tectonophysics 1990; 178: 287-308.
62. Shapira A., and Shamir G. Seismicity Parameters of Seismogenic Zones in and around Israel. The Institute of Petroleum Research and Geophysics 1994. Report No. Z1/567/79 (109).
63. Papazachos BC. and Papaioannou ChA. Long–term Earthquake Prediction in the Aegean area Based on A Time and Magnitude Predicate Model. Pure Applied Geophysics 1993; 140: 595-612.
64. Papioannou ChA. and Papazachos BC. Time–independent and Time–dependent Seismic Hazard in Greece Based on Seismogenic Sources. Bulletin of the Seismological Society of America 2000; 90: 22-33.
65. CMT, Global Centroid Moment Tensor Catalogue: http://www.globalcmt.org/
66. International Seismological Centre, On-line Bulletin, http://www.isc.ac.uk, International Seismological Centre, Thatcham, United Kingdom, 2011.
67. PDE, Preliminary Determination of Epicentre: USGS National Earthquake Information Center (NEIC), http://earthquake.usgs.gov/earthquakes/.
68. RCMT, European-Mediterranean RCMT Catalogue: http://www.bo.ingv.it/RCMT/.
69. Said R. The Geology of Egypt. Elservier, Amsterdam; 1962.
70. Said R. The Geological Evolution of the River Nile. Springer-Verlag, New York Inc., USA; 1981.
71. Said R. The Geology of Egypt. A.A. Balkema, Ralkema, Rptterdam, Brookfield; 1990.
72. Shata A. Structural Development of the Sinai Peninsula (Egypt). Conference Proceedings. 20th International Geological Congress, 1956, Mexico, 1959. p225-249.
73. Neev D. Tectonic Evolution of the Middle East and the Levantine Basin (Easternmost Mediterranean). Geology 1975; 3: 683-686.
74. Neev D., Almagor G., Arad A., Ginzburg A., and Hall J. The Geology of the Southern Mediterranean Sea. GSI, 68. 1976
75. Neev D., and Hall JK. A Global System of Spiraling Geosutures. Journal of Geophysical Research 1982; 87:589-708.
76. El-Shazly EM. The Geology of the Egyptian Region. In: The Ocean Basin and Margins. Volume 4A: The Eastern Mediterranean. Plenum Press. New York-London 1977.
77. Maamoun M. Macroseismic Observation of Principal Earthquakes in Egypt. Bulletin of Helwan Institute of Astronomy and Geophysics. No. 183, 1979.
78. Issawi B. Geology of the South Western Desert of Egypt. Annals of the Geological Survey of Egypt 1981; 2: 57-66.
79. Riad S., EL-Etr HA., and Mokhles A. Basement Tectonics of Northern Egypt as Interpreted from Gravity Data. International Basement Tectonics Association Publication 1983; 4, 209-231.
80. Sestini G. Tectonic and Sedimentary History of NE African Margin (Egypt/Libya). In: JE. Dixon and AHF. Robertson (eds.). The Geological Evolution of the Eastern Mediterranean. Blackwell Scientific Publications, Oxford, 1984. p161-175.

81. Schlumberger. Geology of Egypt. Paper Presented at the Well Evaluation Conference, Schlumberger, Cairo; 1984.

82. Woodward-Clyde Consultants. Earthquake Activity and Stability Evaluation for the Aswan High Dam. Unpublished report. High and Aswan Dam Authority, Ministry of Irrigation, Egypt; 1985.

83. Barazangi M., Seber D., Chaimov T., Best J., Litak R., Al-Saad A and Sawaf T. Tectonic Evolution of the Northern Arabian Plate in Western Syria. In: Boschi E, Mantovani E and Morelli A. (ed.) Recent Evolution and Seismicity of the Mediterranean Region. Kluwer Academic Publishers, the Netherlands; 1993. p117–140.

84. Guiraud RA. and Bosworth W. Phanerozoic Geodynamic Evolution of Northeastern Africa and the Northwestern Arabian Platform. Tectonophysics 1999; 315: 73-108.

85. Abdel Aal A., El Barkooky A., Gerrites M., Meyer H., Schwander M., and Zaki H. Tectonic Evolution of the Eastern Mediterranean Basin and its Significance for Hydrocarbon Prospectivity in the Ultradeep Water of the Nile Delta. The Mediterranean Offshore Conference. Alexandria, Egypt; 2000.

86. Philobbos ER., Riad S., Omran AA. and Othman AB. Stages of Fracture Development Controlling the Evolution of the Nile Valley in Egypt. Egyptian Journal of Geology 2000; 44 (2) 503-532.

87. Hussein IM. and Abd-Allah AM. Tectonic Evolution of the Northeastern Part of the African Continental Margin, Egypt. Journal of African Earth Sciences 2001; 33: 49-68.

88. Drake CL., Girdler RW. A Geophysical Study of the Red Sea. Geophysical Journal of the Royal Astronomical Society 1964; 8: 473-495.

89. Tramontini C. and Davies D. A Seismic Refraction Survey in the Red Sea. Geophysical Journal of the Royal Astronomical Society 1969; 17: 2225–2241.

90. Tealeb A. Depth Determination of Density Contrasts in the Earth's Crust using Autocorrelation Analysis. Bulletin of Academy of Scientific Research, Helwan Institute of Astronomy and Geophysics, Cairo, No. 2008; 1979.

91. Makris J., Stofen B., Maamoun M., Shehata W. Deep Seismic Sounding in Egypt, Part I: The Mediterranean Sea between Crete-Sidi Barani and the Coastal Area of Egypt. Unpublished Report, University of Hamburg ERG, 1979.

92. Makris J., Allam A., Mokhtar T., Basahel A., Dehghani GA. and Bazari M. Crustal Structure in the Northwestern Region of the Arabian Shield and its Transition to the Red Sea. Bulletin of Faculty of Earth Sciences, King Abdulaziz University 1983; 6: 435-447.

93. Makris J., Rihm R. and Allam A. Some Geophysical Aspects of the Evolution and Structure of the Crust in Egypt. In: Greiling SE.-G.a. R.O. (ed.) The Pan-African Belt of Northeast Africa and Adjacent Areas, Tectonic Evolution and Economic Aspects of a Late Proterozoic Orogen: Braunschweig, Friedr. Vieweg and Sohn; 1988. p345–369.

94. Makris J., Henke CH., Egloff F. and Akamaluk T. The Gravity Field of the Red Sea and East Africa. Tectonophysics 1991; 198: 369–381.

95. Rihm R. Seismische Messungen in Roten Meer und ihre interpretation. Diplom Arbeit, Institute fur Geophysik der Univeristat Hamburg, 1984.

96. Gaulier JM., Le Pichon X., Lyberis N., Avedik F., Geli L., Moretti I., Deschamps A. and Hafez S. Seismic Study of the Crust of the Northern Red Sea and Gulf of Suez. Tectonophysics 1988; 153: 55–88.

97. Rihm R., Makris J., Moller L. Seismic Surveys in the Northern Red Sea: Asymmetric Crustal Structure. Tectonophysics 1991; 198: 279–295.

98. Shaaban MA., El Eraqi MA., Mamdouh EM. Deep Tectonics of Northern Eastern Desert of Egypt as Integrated from Gravity and Seismic Data. Journal of King Abdulaziz University: Earth Sciences 1994; 7: 75–88.

99. Abd El-Hafiez H. The Role of Earthquake Analysis for Modeling the Dahshour Area Egypt. MSc. thesis. Ain Shams University, Egypt; 1996.

100. Dorre AS., Carrara E., Cella F., Grimaldi M., Hady YA., Hassan H., Rapolla A. and Roberti N. Journal of African Earth Sciences 1997; 25: 425-434.

101. Seber D., Steer D., Sandvol E., Sandvol C., Brindisi C., and Barazangi M. Design and Development of Information Systems for the Geosciences: An Application to the Middle East. GeoArabia 2000; 5 (2) 269-296.

102. Mohamed H. and Miyashita K. One-dimensional Velocity Structure in the Northern Red Sea Area, Deduced from Travel Time Data. Earth's Planet Spaces 2001; 53: 695–702.

103. El-Khrepy S. Tomographic Modeling of Dahshour Area Local Earthquakes, Northern Egypt. MSc. thesis. Ain Shams University, Egypt, 2001.

104. El-Khrepy S. Detailed Study of the Seismic Waves Velocity and Attenuation Models Using Local Earthquakes in the Northeastern Part of Egypt. PhD. thesis. Mansoura University, Egypt, 2008.

105. Koulakov I. and Sobolev SV. Moho Depth and Three-dimensional P and S Structure of the Crust and Uppermost Mantle in the Eastern Mediterranean and Middle East Derived from Tomographic Inversion of Local ISC Data. Geophysical Journal International 2006; 164: 218-235.

106. Gharib A. Crustal Structure of Tushka Region, Abu-Simbel, Egypt, Inferred from Spectral Ratios of P-waves of Local Earthquakes. Acta Geophysica 2006; 54: 361- 377.

107. Salah MK. Crustal Structure Beneath Kottamiya Broadband Station, Northern Egypt from Analysis of Teleseismic Receiver Functions. Journal of African Earth Sciences 2011; 60: 353–362.

108. Abdelwahed MF., El-Khrepy S., and Qaddah A. Three-dimensional Structure of Conrad and Moho Discontinuities in Egypt. Journal of African Earth Sciences 2013; 85: 87-102.

109. Araya R., and Der Kiureghian A. Seismic Hazard Analysis: Improved Models, Uncertainties and Sensitivities. Report to the National Science Foundation, Earthquake Enginnering Research Center 1988. Report No. UCB/EERC-90/11.

110. Gutenberg B. and Richter CF. Frequency of Earthquakes in California. Bulletin of the Seismological Society of America1944; 34: 185-188.

111. Salamon A., Hofstetter A., Garfunkel Z. and Ron H. Seismicity of the Eastern Mediterranean Region: Perspective from the Sinai Subplate. Tectonophysics 1996; 263: 293-305.

112. Salamon A., Hofstetter A., Garfunkel Z. and Ron H. Seismotectonics of the Sinai Subplate-the Eastern Mediterranean Region. Geophysical Journal International 2003; 155: 149-173.

113. Garfunkel Z., Zak I. and Freund R. Active Faulting in the Dead Sea Rift. Tectonophysics 1981; 80: 1–26.

114. Reches Z. and Hoexter DF. Holocene Seismic Activity in the Dead Sea Area. Tectonophysics 1981; 80: 235-254.

115. Marco S., Agnon A., Ellenblum R., Eidekman A., Basson U. and Boas A. 817-year-old Walls Offset Sinistrally 2.1 m by the Dead Sea Transform, Israel. Journal of Geodynamics 1997; 24:11–20.

116. Marco S., Heimann A., Rockwell KT. and Agnon A. Late Holocene Earthquake Deformations in the Jordan Gorge Fault, Dead Sea Transform. In Abstracts of Israel Geological Society Annual Meeting. Ma'alot, 2000, p85.

117. Zilberman E., Amit R., Heimann A. and Porat N. Changes in Holocene Paleoseismic Activity in the Hula Pull-apart Basin, Dead Sea Rift, Northern Israel. Tectonophysics 2000; 321: 237–252.

118. Amit R., Zilberman E., Porat N., Enzel Y. Relief Inversion in the Avrona Playa as Evidence of Large-magnitude Historical Earthquakes, Southern Arava Valley, Dead Sea Rift. Quaternary Research 1999; 52 (1) 76 – 91.

119. Zhang H. and Niemi TM. Slip Rate of the Northern Wadi Araba Fault, Dead Sea Transform, Jordan. GSA Abstracts with Programs 1999; 31, A-114.

120. Klinger Y., Avouac J., Abou Karaki N., Dorbath L., Bourles and Reyss J. Slip Rate on the Dead Sea Transform Fault in Northern Araba Valley, Jordan. Geophysical Journal International 2000; 142: 755-768.

121. Gomez F., Meghraoui M., Darkal AN., Sbeinati R., Darawcheh R., Tabet C., Knawlie M., Charabe M., Khair K. and Barazangi M. Coseismic Displacements along the Serghaya Fault: An Active Branch of the Dead Sea Fault System in Syria and Lebanon. Journal of the Geological Society of London 2001; 158: 405-408.

122. Garfunkel Z. Internal Structure of the Dead Sea Leaky Transform (Rift) in Relation to Plate Kinematics. Tectonophysics 1981; 80, 81-108.

123. Walley CD. A Braided Strike-slip Model for the Northern Continuation of the Dead Sea Fault, and its Implications to Levantine Tectonics. Tectonophysics 1988; 145: 63–72.

124. Girdler RW. The Dead Sea Transform Fault System. Tectonophysics 1990; 180: 1–13.

125. Gomez F., Meghraoui M., Darkal AN., Hijazi F., Mouty M., Suleiman Y., Sbeinati R., Darawcheh R., Al-Ghazzi R., Barazangi M. Holocene Faulting and Earthquake Recurrence along the Serghaya Branch of the Dead Sea Fault System in Syria and Lebanon. Geophysical Journal International 2003; 153: 1–17.

126. Meghraoui M., Cakir Z., Masson F., Mahmoud Y., Ergintav S., Alchalbi A., Inan S., Daoud M., Yonlu O., Altunel E. Kinematic Modelling at the Triple Junction between the Anatolian, Arabian, African Plates (NW Syria and in SE Turkey). Geophysical Research Abstracts 2011; 13, EGU2011-12599, EGU General Assembly, Vienna.

127. Karabacak V., and Altunel E. Evolution of the Northern Dead Sea Fault Zone in Southern Turkey. Journal of Geodynamics 2013; 65: 282–291.

128. Mahmoud Y., Masson F., Meghraoui M., Cakir Z., Alchalbi A., Yavasoglu H., Yönlü O., Daoud M., Ergintav S., Inan S. Kinematic Study at the Junction of the East Anatolian Fault and the Dead Sea Fault from GPS Measurements. Journal of Geodynamics 2013; 67: 30–39.

129. Dziewonski AM., Ekstrom G., Salganik MP. Centroid Moment Tensor Solutions for October – December 1995. Physics of the Earth and Planetary Interiors 1997; 101: 1 – 12.

130. Garfunkel Z. The Tectonics of the Western Margins of the South Arava. PhD. thesis. The Hebrew University of Jerusalem, Israel, 1974.

131. Ben-Avraham Z., Almagor G. and Garfunkel Z. Sediments and Structure of the Gulf of Elat (Aqaba) Northern Red Sea. Sedimentary Geology 1979; 23: 239-267.

132. Eyal M, Eyal Y., Bartov Y. and Steinitz G. The Tectonic Development of the Western Margin of the Gulf of Eilat (Aqaba) Rift. Tectonophysics 1981, 80: 39-66.

133. Shapira A., Jarradat M. Earthquake Risk and Loss Assessment in Aqaba and Eilat Regions. Submitted to the US Aid-Merc Program 1995.

134. Heidbach O., and Ben-Avraham Z. Stress Evolution and Seismic Hazard of the Dead Sea Fault System. Earth and Planetary Science Letters 2007; 257: (1-2) 299-312.

135. Pinar A. and Türkelli N. Source Inversion of the 1993 and 1995 Gulf of Aqaba Earthquakes. Tectonophysics 1997; 283: 279–288.

136. Klinger Y., Rivera L., Haessler H. and Maurin JC. Active Faulting in the Gulf of Aqaba: New Knowledge from the MW 7.3 Earthquake of 22 November 1995. Bulletin of Seismological Society of America 1999; 89: 1025–1036.

137. Hofstetter A., Thio HK. and Shamir G. Source Mechanism of the 22/11/1995 Gulf of Aqaba Earthquake and its Aftershock Sequence. Journal of Seismology 2003; 7: 99-114.

138. Abdel Fattah AK., Hussein HM. and El Hady S. Another Look at the 1993 and 1995 Gulf of Aqaba Earthquake from the Analysis of Teleseismic Waveforms. Acta Geophysica 2006; 54 (3) 260-279.

139. Lyberis N. Tectonic Evolution of the Gulf of Suez and the Gulf of Aqaba. Tectonophyics 1988; 153: 209-220.

140. Bayer H., Hötzl H., Jado A., Bocher B. and Voggenreiter W. Sedimentary and Structural Evolution of the Northwest Arabian Red Sea Margin. Tectonophysics 1988; 153: 137-151.

141. Gerson R., Grossman S., Amit R. and Greenbaum N. Indicators of Faulting Events and Periods of Quiescence in Desert Alluvial Fans. Earth Surface Processes and Landforms 1993; 18: 181-202.

142. Steinitz G., Bartov Y., and Hunziker JC. K-Ar Age Determinations of Some Miocene-Pliocene Basalts in Israel: Their Significance to the Tectonics of the Rift Valley. Geological Magazine 1978; 115: 329–340.

143. Moustafa AR. and Khalil MH. Superposed Deformation in the Northern Suez Rift, Egypt: Relevance to Hydrocarbon Exploration. Journal of Petroleum Geology 1995; 18: 245–266.

144. Bartov Y. A Structural and Paleographic Study of the Central Sinai Faults and Domes. PhD. thesis (in Hebrew with an English Abstract). The Hebrew University of Jerusalem, Israel; 1974.

145. Moustafa AR. and Khalil MH. Rejuvenation of the Eastern Mediterranean Passive Continental Margin in Northern and Central Sinai: New Data from the Themed Fault. Geological Magazine 1994; 131: 435-448.

146. El-Isa Z., Merghelani H. and Bazari M. The Gulf of Aqaba Earthquake Swarm of 1983. Geophysical Journal of the Royal Astronomical Society 1984; 76: 711-722.

147. Ben-Menahem A., Nur A., and Vered M. Tectonics, Seismicity and Structure of the Afro-Eurasian Junction--The Breaking of an Incoherent Plate. Physics of the Earth and Planetary Interiors 1976; 12: 1-50.

148. Gardosh M., Reches Z. and Garfunkel Z. Holocene Tectonic Deformation along the Western Margins of the Dead Sea. Tectonophysics 1990; 180: 132-137.

149. Heimann A. The Development of the Dead Sea Rift and its Margins in Northern Israel in the Pliocene and Pleistocene. PhD. thesis. The Hebrew University of Jerusalem, Israel; 1990.

150. Shamir G., Bartov A., Fleischer L., Arad V., Rosensaft M. Preliminary Seismic Zonation. Geological Survey of Israel. Report No. GSI/12/2001, Geophysical Institute of Israel Report No. GII 550/ 95/01/ (1); 2001.

151. Cochran JR. A Model for the Development of the Red Sea. American Association of Petroleum Geologists Bulletin 1983; 67: 41-69.

152. Cochran JR., Martinez F., Steckler and Hobart MA. Conrad Deep, a New Northern Red Sea Deep, Origin and Implications for Continental Rifting. Earth and Planetary Science Letters 1986; 78: 18-32.

153. Girdler RW and Styles P. Two Stage Sea-floor Spreading. Nature 1974; 247: 7-11.

154. Roeser HA. A Detailed Magnetic Survey of the Southern Red Sea. Geologie Jahrbuch 1975; 13: 131–153.

155. LaBrecque JL., and Zitellini N. Continuous Sea Floor Spreading in the Red Sea: An Alternative Interpretation of Magnetic Anomaly Pattern. The American Association of Petroleum Geologists Bulletin 1985; 4: 513-524.

156. Makris J. and Rihm R. Shear-controlled Evolution of the Red Sea: Pull-Apart Model. Tectonophysics 1991; 198: 441–466.

157. Searle RC. and Ross DA. A Geophysical Study of the Red Sea Axial Trough Between 20.58 and 22.8N. Geophysical Journal of the Royal Astronomical Society 1975; 43: 555–572.

158. Ghebreab W. Tectonics of the Red Sea Region: Reassessed. Earth-Science Reviews 1998; 45: 1-44.

159. Cochran JR. and Martinez F. Evidence from the Northern Red Sea on the Transition from Continental to Oceanic Rifting. Tectonophysics 1988; 153: 25-53.

160. Bonati E. Punctiform Initiation of Seafloor Spreading in the Red Sea during Transition from a Continental to an Oceanic Rift. Nature 1985; 316: 7-33.

161. Daggett P., Morgan P., Boulous F., Hennin S., El-Sherif A., El-Sayed A., Basta N. and Melek Y. Seismicity and Active Tectonics of the Egyptian Red Sea Margin and the Northern Red Sea. Tectonophysics 1986; 125: 313-324.

162. McKenzie DP., Davies D. and Molnar P. Plate Tectonics of the Red Sea and East Africa. Nature 1970; 226: 243–248.

163. Robson D. The Structure of the Gulf of Suez (Clysmic) Rift with Special Reference to the Eastern Side. Geological Society of London 1971; 115: 247-276.

164. Garfunkel Z. and Bartov Y. The Tectonics of the Suez Rift. Geological Survey of Israel Bulletin 1977.

165. Jackson J., White N., Garfunkel Z. and Anderson H. Relation between Normal-fault Geometry, Tilting and Vertical Motions in Extensional Terrains: An Example from the Southern Gulf of Suez. Journal of Structural Geology 1988; 10 (2) 155-170.

166. Bosworth W. and Taviani M. Late Quaternary Reorientation of Stress Field and the Extension Direction in the Southern Gulf of Suez, Egypt: Evidence from Uplifted Coral Terraces, Mesoscopic Fault Arrays and Borehole Breakouts. Tectonics 1996; 15: 791-802.

167. Moustafa AR. Internal Structure and Deformation of an Accommodation Zone in the Northern Part of the Suez Rift. Journal of Structural Geology 1996; 18: 93-107.

168. Moustafa AM. Block Faulting in the Gulf of Suez. Conference Proceedings, 1976, Cairo, Egypt. 5th Egyptian General Petroleum Corporation Exploration Seminar, 35 p.

169. Bosworth W. A High-strain Rift Model for the Southern Gulf of Suez, Egypt. In: Lambiase L L (ed.) Hydrocarbon Habitat in Rift Basins. Geological Society of London 1985. Special Publication 80, p75–102.

170. Moustafa AR. and Fouda HG. Gebel Sufr El Dara Accommodation Zone, Southwestern Part of the Suez rift. Middle East Research Center, Ain Shams University, Earth Science Series 1988; 2: 227–239.

171. Chorowicz J. and Sorlien C. Oblique Extensional Tectonics in the Malawi Rift, Africa. Geological Society of America Bulletin 1992; 104: 1015–1023.

172. Maler MO. Dead Horse Graben: A West Texas Accommodation Zone. Tectonics 1990; 9: 1257–1268.

173. Boccaletti M., Getaneh A., and Tortoorici L. The Main Ethiopian Rift: An Example of Oblique Rifting. Annales Tectonicae 1992; 6: 20–25.

174. Lacombe O., Angelier J., Byrne D., and Dupin JM. Eocene-Oligocene Tectonics and Kinematics of the Rhine-Saone Continental Transform Zone (Eastern France). Tectonics 1993; 12: 874-888.

175. Colletta B., Le Quellec P., Letouzey J., and Moretti I. Longitudinal Evolution of the Suez Rift Structure, Egypt. Tectonophysics 1988; 153: 221–233.

176. Le Pichon X. and Gaulier J. The Rotation of the Arabia and Levant Fault System. Tectonophysics 1988; 153: 271-294.

177. Mart Y. The Dead Sea Rift: From Continental Rift to Incipient Ocean. Tectonophysics 1991; 197:155–179.

178. Ben-Menahem A. Earthquake Catalogue for the Middle East (92 BC-1980 AD). Bollettino di Geofisica Teorica e Applicata 1979; 21: 245-310.

179. Ben-Menahem A. and Aboodi E. Tectonic Pattern in the Northern Red Sea Region. Journal of Geophysical Research 1971; 76: 2674-2689.

180. Younes AI., and McClay K. Development of Accommodation Zones in the Gulf of Suez-Red Sea Rift, Egypt. The American Association of Petroleum Geologists Bulletin 2002; 86: 1003-1026.

181. Fairhead JD. and Girdler RW. The Seismicity of the Red Sea, Gulf of Aden and Afar Triangle. Philosophical Transactions of the Royal Society of London 1970; 267: 49-74.

182. Maamoun M. and El Khashab HM. Seismic Studies of the Shedwan (Red Sea) Earthquake. Helwan Institute of Astronomy and Geophysics. No. 171, 1978.

183. Jackson JA. and McKenzei DP. Active Tectonics of the Alpine Himalayan Belt between Western Turkey and Pakistan. Geophysical Journal of the Royal Astronomical Society 1984; 77: 185-264.

184. Huang P. and Soloman S. Centroid Depth and Mechanisms of Mid-ocean Ridge. Journal of Geological Research 1987; 92: 1361-1383.

185. Badawy A., and Horváth F. Sinai Subplate and Kinematic Evolution of the Northern Red Sea. Journal of Geodynamics 1999a; 27: 433-450.

186. Badawy A., and Horváth F. Seismicity of the Sinai Subplate Region: Kinematic Implications. Journal of Geodynamics 1999b; 27: 451-468.

187. Badawy A., and Horváth F. Recent Stress Field of the Sinai Subplate Region. Tectonophysics 1999c; 304: 385-403.

188. Badawy A. Status of the Crustal Stress as Inferred from Earthquake Focal Mechanisms and Borehole Breakouts in Egypt. Tectonophysics 2001; 343 (1-2) 49-61.

189. El-Gaby S. Architecture of the Egyptian Basement Complex. 5th International Conference on Basement Tectonics, Egypt; 1983.

190. El-Gaby S., List FK., and Tehrani R. Geology, Evolution and Metallogenesis of the Pan-African Belt in Egypt. In: S.El-Gaby and R. O. Greiling (eds), The Pan-African Belt of Northeast African and Adjacent Areas, Fried. Vieweg and Shon, Braun Schweig, Wiesbaden 1988, p17-68.

191. Stern RJ., and Hedge CE. Geochronologic and Isotopic Constraints on Late Precambrian Crustal Evolution in the Eastern Desert of Egypt. American Journal of Science 1985; 285: 97-127.

192. Issawi B. The Geology of Kurkur-Dungul Area. General Egyptian Organization for Geological Research and Mining; Cairo, Egypt. Geological Survey. No. 46, 101 pp; 1969.

193. Issawi B. Geology of the Southwestern Desert of Egypt. Annals of the Geological Survey of Egypt 1982, 215 p.

194. Issawi B. Geology of the Aswan Desert. Annals of the Geological Survey of Egypt; 1987.

195. Fat-Helbary RE. A Study of the Local Earthquake Magnitude Determination Recorded by Aswan Seismic Network. MSc. thesis. Assiut University, Sohag Branch, Egypt; 1989.

196. El-Younsy ARM., Ibrahim HA., Senosy MM. and Galal WF. Structural Characteristics and Tectonic Evolution of the Area around the Qena Bend,

Middle Egypt. 6th International Conference on the Geology of Africa, 2010, Assiut, Egypt.

197. Beadnell HJL. Dakhla Oasis, Its Topography and Geology: Egypt. Survey Department 1901; 104 Pages, 9 Maps, 7 Figures.

198. Sandford KS. Paleolithic Man and the Nile Valley in Upper and Middle Egypt. Chicago University Oriental Institute Publications 1934; 18: 1-131.

199. El-Gamili M. A Geophysical Interpretation of A part of the Nile Valley, Egypt Based on Gravity Data. Journal of Geology 1982. Special Volume, Part 2: 101-120.

200. Abdel-Rahman MA., and El-Etr HA. The Orientational Characteristics of the Structure Grain of the Eastern Desert of Egypt. In: Symposium of the Evolution and Mineralization of the Arabian-Nubian Shield. Institute of Applied Geology, Jeddah, Saudi Arabia; 1978.

201. Abdel Tawab S., Helal A., Deweidar H. and El-Sayed A. Surface Tectonic Features of 12 Oct., 1992 Earthquake, Egypt, at the Epicentral area. Ain Shams Scientific Bulletin 1993; Special Issue, p124-136.

202. Maamoun M., Megahed A., Hussein A. and Marzouk I. Preliminary Studies on Dahashour Earthquake. National Research Institute of Astronomy and Geophysics, Cairo, Egypt. (Abstract), 1993.

203. Mousa HH. Earthquake Activity in Egypt and Adjacent Regions and its Relation to the Geotectonic Features. MSc. thesis. Mansoura University, Egypt, 1989.

204. Hassib GH. A Study of Focal Mechanism for Recent Earthquakes in Egypt and their Tectonic Implication. MSc thesis. Assiut University, Egypt, 1990.

205. Megahed A. and Dessoky MM. The Ismailia (Egypt) Earthquake of January 2nd, 1987 (Location, Macroseismic Survey, Radiation Pattern of First Motion and its Tectonic Implications), 1988.

206. Hassoup A. and Tealab A. Attenuation of Intensity in the Northern Part of Egypt Associated with the May 28, 1998 Mediterranean Earthquake. Acta Geophysica Polonica 2000; 48: 79-92.

207. Riad S. and Hosney H. Fault Plane Solution for the Gilf Kebir Earthquake and the Tectonics of the Southern Part of the Western Desert of Egypt. Annals of the Geological Survey of Egypt 1992; 18: 239-248.

208. Sykes LR. Intraplate Seismicity, Reactivation of Preexisting Zones of Weakness, Alkaine Magmatism and Other Tectonic Postdating Continental Fragmentation. Reviews of Geophysics and Space Physics 1987; 16: 621-688.

209. Talwani P. and Rajendrank. Some Seismological and Geometric Features of Intraplate Earthquakes. Seismological Research Letters 1991; 59, 305-310.

210. McKenzie D. Plate Tectonics of the Mediterranean Region. Nature1970; 326: 239-243.

211. McKenzie D. Active Tectonics in the Mediterranean Region. Geophysical Journal of the Royal Astronomical Society 1972; 30: 109-185.

212. Dewey JF., Pitman WC., Ryan WBF. and Bonnin J. Plate Tectonics and the Evolution of the Alpine System. Bulletin of the Geological Society of America 1973; 84: 3137-3180.

213. Westaway R. Present-day Kinematics of the Middle East and Eastern Mediterranean. Journal of Geophysical Research 1994; 99 (6) 12071-12090.

214. Kiratzi A. and Papazachos C. Active Crustal Deformation from the Azores Triple Junction to the Middle East. Tectonophysics 1995; 243: 1-24.

215. Ambraseys N. and Jackson J. Faulting Associated with Historical and Recent Earthquakes in the Eastern Mediterranean Region. Geophysical Journal International 1998; 133: 390-406.

216. McClusky S., Balassanian S., Barka A., Demir C., Ergintav S., Georgiev I., Gurkan O., Hamburger M., Hurst K., Kahle H., Kastens K., Kekelidze G., King R., Kotzev V., Lenk O., Mahmoud S., Mishin A., Nadariya M., Ouzounis A., Paradissis D., Peter Y., Prilepin M., Reilinger R., Sanli I., Seeger H., Tealeb A., Toksoz MN. and Veis G. Global Positioning System Constraints on Plate Kinematics and Dynamics in the Eastern Mediterranean and Caucasus. Journal of Geophysical Research 2000; 105 (3) 5695–5719.

217. Vidal N., Alvarez-Marron J. and Klaeschen D. Internal Configuration of the Levantine from Seismic Reflection Data (Eastern Mediterranean). Earth and Planetary Science Letters 2000a; 180: 77–89.

218. Vidal N., Alvarez-Marron J. and Klaeschen D. The Structure of the Africa–Anatolia Plate Boundary in the Eastern Mediterranean. Tectonics 2000b, 19: 723–739.

219. Vidal N., Klaeschen D., Kopf A., Docherty C., Von-Huene R. and Krasheninnikov VA. Seismic Images at the Convergence Zone from South of Cyprus to the Syrian Coast, Eastern Mediterranean. Tectonophysics 2000c; 329: 157–170.

220. Papazachos BC. and Papaioannou ChA. Lithospheric Boundaries and Plate Motions in the Cyprus Area. Tectonophysics 1999; 308: 193–204.

221. Ziegler PA. Evolution of the Arctic-North Atlantic and Western Tethys. The American Association of Petroleum Geologists Memoir 1988; 43, 198p.

222. Meulenkamp JE., Wortel MJR., van Wamel WA., Spakman W. and Hoogerduyn Strating E. On the Hellenic Subduction Zone and the Geodynamic Evolution of Crete since the Late Middle Miocene. Tectonophysics 1988; 146: 203–215.

223. Dewey JF., Helman ML., Turco E., Hutton DHW., and Knott SD. Kinematics of the Western Mediterranean. In: Alpine Tectonics, edited by M.P. Coward, D. Detrich and R.G. Park, Geological Society of London, Special Publication 1989, 45: 265-283.

224. SHARE "Seismic Hazard Harmonization in Europe", 2013. http://www.share-eu.org/.

225. Woessner J., Giardini D., and the SHARE consortium. Seismic Hazard Estimates for the Euro-Mediterranean Region: A Community-based Probabilistic Seismic Hazard Assessment. Proceedings of the 15th World Conference of Earthquake Engineering, Lisbon, Portugal, 2012.

226. Papioannou ChA. A Model for the Shallow and Intermediate-depth Seismic Sources in the Eastern Mediterranean Region. Bollettino di Geofisica 2001; 42: 57-73.

227. Demircioglu MB., Sesetyan K., Durukal E. and Erdik M. Assessement of Earthquake Hazard in Turkey. Conference Proceedings, 4th International Conference on Earthquake Geotechnical Engineering, 25-28 June 2007, Thessaloniki, Greece. Springer, New York.

228. Demircioglu MB. The Earthquake Hazard and Risk Assessment for Turkey. PhD. thesis. Bogazici University, Turkey, 2010.
229. Papazachos BC., and Comninakis PE. Geophysical Features of the Greek Island Arc and Eastern Mediterranean Ridge. Final Proceedings. Seances de la Conference Reunie a Madrid, 1969/1970. Madrid, Spain. 16: 74-75.
230. Papazachos BC., and Comninakis PE. Geophysical and Tectonic Features of the Aegean Arc. Journal of Geophysical Research 1971; 76: 8517-8533.
231. Comninakis PE., and Papazachos BC. Space and Time Distribution of the Intermediate Depth Earthquakes in the Hellenic Arc. Tectonophysics 1980; 70: 35-47.
232. Papazachos BC., Papadimitriou EE., Karakostas BG. and Karakaisis GF. Long-term Prediction of Great Intermediate-Depth Earthquakes in Greece. Proceedings of the 12th Regional Seminar on Earthquake Engineering, EAEE-EPPO, Halkidiki, 1985, 1-12.

CITATION

R. Sawires, J.A. Peláez, R.E. Fat-Helbary, H.A. Ibrahim and M.T. García Hernández (2015). An Updated Seismic Source Model for Egypt, Earthquake Engineering - From Engineering Seismology to Optimal Seismic Design of Engineering Structures, Prof. Abbas Moustafa (Ed.), ISBN: 978-953-51-2039-1, InTech, DOI: 10.5772/58971.

CHAPTER 3

Earthquakes and Dams

Hasan Tosun[1]

[1] Civil Engineering Department, Uşak University, Uşak, Turkey

INTRODUCTION

Earthquake is defined as a sudden and rapid shaking of the earth caused by the breaking and shifting of rock beneath the Earth's surface and it creates seismic waves, which can result in damages and failures on man-made structures constructed on the crust of earth [18]. Dams and large reservoirs constructed on the area with high seismicity, pose a high-risk potential for downstream life and property. It is clear that active faults, which are located close to dam sites, can induce to damaging deformation of the embankment as based on instability of the dam and strength loss of foundation materials. Scientists have realized so many researches for explaining the behavior of earth structures under seismic forces.

Earthquake effects on dams mainly depend on dam types. [28] stated that safety concerns for embankment dams subjected to earthquakes involve either the loss of stability due to a loss of strength of the embankment and foundation materials or excessive deformations such as slumping, settlement, cracking and planer or rotational slope failures. According to [9], safety requirements for concrete dams subjected to dynamic loadings should involve evaluation of the overall stability of the structure, such as verifying its ability to resist induced lateral forces and moments and preventing excessive cracking of the concrete.

Earthquakes can result in damages or failures for dam structures, while dams with large reservoirs can induce to earthquakes. Case studies about the seismic performance of dams under large earthquakes are available in the literature. [31] state that earthquake safety of dams is an important phenomenon in dam engineering and requires more comprehensive seismic studies for understanding the seismic behavior of dams subjected to severe earthquakes. It is a well-known phenomenon that earthquakes can result damages and failures for dams and their appurtenant structures. There is another fact that dams with large reservoirs also trigger earthquake.

Ground shaking from earthquakes can collapse dams. There are some important cases, which subjected to damages and failures after earthquake. Lower San Fernando Dam in USA is first example failed as a result of liquefaction phenomenon under the earthquake loading conditions. In case of the May 12, 2008 Wenchuan earthquake in China many dams and reservoirs had been subjected to strong ground shaking. So many dams and hydropower plants were damaged. During the 2001 Bhuj earthquake in Gujarat, India, 245 dams had been affected and rehabilitated or strengthened after the earthquake. Also, in the case of the March 11, 2011 Tohoku earthquake in Japan, damages were observed about 400 dams and the 18 m high embankment dam failed and 8 people lost their live.

Large reservoirs can trigger earthquake. According to recent surveys, Reservoir Triggering Seismicity (RTS) has been observed at over 100 locations worldwide [4, 19, 20]. The largest and most damaging earthquake triggered by a man-made reservoir may be the 7.9-magnitude Sichuan earthquake in May 12, 2008. One of the most serious cases was in 1967 in Koyna, India. The magnitude of this earthquake was 6.3. Also significant effects have been observed Hsingfengkiang dam in China, Kariba dam in Zimbabwe and Kremasta dam in Greece. The effect of reservoir loading on the existing stress field has been investigated by several studies [1, 5, 13, 14, 15, 19, 20, 21, 22, 23]. The field studies indicates that the main factors acting reservoir seismicity are in-situ stress conditions, availability of fractures and faults, geology of the regional area, dimensions of the reservoir and the nature of reservoir level fluctuations.

The paper gives an overview on the dams, which are under the effects of strong ground motions. It investigates the effects of earthquake on dams, also effects of dam on earthquake occurrence. Some cases are given to explain both phenomena and clarify the total risk of large dam structures when considered earthquake effects. The subjects presented in the paper were addressed by the international committees and recent surveys. It mentions main requirements for large dams on view of earthquake engineering to find rational design solutions. The purpose of this paper is to sketch the state of the art in dam engineering, as based on lessons learnt from seismic events.

EFFECTS OF EARTHQUAKE ON DAMS

Damages to dams and their appurtenant facilities may result from (1) direct fault movement across the dam foundation or (2) from ground motion induced at the dam site by an earthquake located at some distance from the dam. The second one is commonly seen, however first one results to more serious problems for dams and their appurtenant structures. A good example to damages resulted by ground shaking vibrations in dams is Sefid buttress dam, which was damaged near crest due to ground shaking the 1990 Manjil earthquake with a magnitude of 7.5 in Iran. In this dam, damages have been observed near crest due to ground vibration. For fault movements in dam site, the Shih-Kang weir can be considered as good case study. In this dam, two openings were failed due to large movements of Chelungpu fault during the magnitude of 7.3 in Chi-Chi earthquake of September 1999 in Tawian. After severe damages observed on this dam, dam engineers more seriously considered active or seismogenic faults on dam sites. Because dams located on active faults pose significant risk for total stability of project and public safety.

Liquefaction is defined as a phenomena in which the strength and stiffness of a saturated soil is reduced earthquake shaking. It generally means the state change from solid to liquid. Lower San Fernando Dam in USA is first known dam failed as a result of liquefaction phenomenon under the earthquake loading conditions. During the 1972 San Fernando Earthquake, Lower San Fernando Dam failed a result of liquefaction phenomena [12, 17, 24]. Its embankment with the structures on crest slid into the reservoir. In other words, approximately 3.0 million cubic meter of dam embankment was displaced into the reservoir. The 1994 Northridge earthquake, some ground movement with minor cracking seems to have occurred at the sites of Los Angeles Dam, which was constructed to replace the San Fernando Reservoir. There was significant differential settlement of the ground of about 5 cm in the northern section, and 20 cm in the southwestern section of the site [24].

During the 2001 Bhuj earthquake in Gujarat, India, 245 dams had been affected and rehabilitated or strengthened after the earthquake [34]. Due to Mid Niigata Prefecture Earthquake in 2004, Japan, several embankment dams and some off-stream impounding facilities for power generation and irrigation system suffered damages such as cracks on dam bodies [10].

In case of the May 12, 2008 Wenchuan earthquake in China many structures about 1803 dams and 403 hydropower plants having a total installed capacity of 3.3 GW were damaged due to strong ground shaking. Most of dams were small earth dams with exception of four large dams having a height greater than 100 m. According to Chinese officials the earthquake occurred along the Longmenshan fault, which is a thrust structure along the border of the Indo-Australian Plate and Eurasian Plate, the rupture lasted 120 sec, the rupture propagated at an average speed of 3.1 km/s toward northeast. The rupture length and focus depth is about 300 km and 10 km, respectively. The maximum displacement was recorded as 9.0 m [6]. As a result of this earthquakes so many elements of dam such as dam body, spillways, powerhouses, penstocks, switchyards, hydro-mechanical and electro-mechanical equipments, temporary structures were damaged, other disasters such as rockfalls, landslides and landslide dams were observed. No dams were failed during this earthquake, although there were so many damaged dams. [8] states that dams must be designed to withstand strong earthquakes, which can seriously result multiple hazards.

In the case of the March 11, 2011 Tohoku earthquake in Japan, damages were observed about 400 dams and the 18 m high embankment dam failed and 8 people lost their live [34].

The dams, which are located on shear zones, have high risk potential when they are subjected to strong ground motion. There are some examples in India for structures located at the northern India. One of the Namada Valley dam, which is built at the triple junction of the fault zones, tectonically and geologically a disturbed area. Terhi dam in India has also similar position under dynamic loading conditions. Researchers states that Terhi dam might release energy along the fault segment between Nepal and Tibet and also trigger an earthquake which has a magnitude close to or greater than 8.0.

In Turkey, there is a sheared zone which is close the triple junction of the famous strike slip faults in east of Turkey. [28] stated that Surgu dam, which damaged on the Dogansehir earthquake with Ms of 5.8 in 1986, Polat dam and Cat dam have the PGA values of 0.256g, 0.170g and 0.211g, respectively. The geology of dam sites are very complicated and frequently jointed, fractured and faulted. The author points out the fact that these dams are under the influence of local near-source zone and have high-risk potential for earthquake conditions. The author's thought was absolutely confirmed by damage on the Dogansehir earthquake with Ms of 5.8 on Surgu dam.

In general, strong ground shaking can result in the instability of the embankment and loss of strength at the foundations [2, 9, 16, 17]. Most of dam engineers have thought that embankment dams are suitable types when well compacted according to the specification, However, it is not an acceptable thought that embankment dams can be induced to damages and failures even if well compacted, while they are under near source effect.

There is no one major problem in seismic safety of embankment dams. Whereas near source effect seems the most serious problem for embankment dams. [28] reveals the fact that active faults, which are very close to the foundation of dams, have the potential to cause damaging displacement of the structure. Especially Concrete Faced Rockfill Dams (CFRD's) have high risk potential when considered near source effect (earthquake epicenter to dam axis is less than 10 km). This phenomenon is dealed with the fact that the transferred energy by rockfill is not absorbed by concrete face during earthquake. Wieland (2010) state that until the Wenchuan earthquake of 12 May 2008 no large concrete face rockfill dam (CFRD) was subjected to strong ground shaking. He questioned that faced concrete of CFRD's can have a behavior as the river embankment which was subjected the 21 September 1999 Chi-Chi earthquake in Taiwan [32 and 33]. Figure 1 shows buckling of river embankment lining after the earthquake.

EFFECTS OF DAMS ON SEISMICITY OF THE REGION

Large reservoirs can trigger earthquake. This phenomenon is defined as Reservoir-induced Seismicity that is mainly depended to excessive water pressure created in the micro-cracks and fissures in the foundation units under and near the reservoir. Water within the rock masses under huge hydrostatic pressure acts to lubricate faults, which are already under tectonic strain, however are prevented from slipping by friction of rock planes. It is clearly known that it mainly depends on nature of structural geology and lithology of surrounding rocks. However, it is very difficult to accurately predict when and where reservoir induced earthquake will occur. ICOLD recommends that Reservoir Triggered Seismicity (RTS) should be considered for reservoirs having a depth more than 100 m. USCOLD has reported that Reservoir Induced Seismicity (RIS) should be taken into account for reservoirs deeper than 80-100m.

Figure 1: Buckling of river embankment lining after the 1999 Chi-Chi earthquake [32].

It is clear that number of seismic events increases near reservoir areas of large dams after impounding sequence. The earthquake seismicity was firstly observed in 1929 for Marathon dam having 60 m height, Greece. Increase in seismicity was also seen in 1935 after the impounding of Hoover dam, which is a concrete arch dam with a height of 220 m. Up to now, RTS has been observed on over 100 dams in the world. The earthquake intensity has increased after impounding of Keban Dam, which is the second largest dam of Turkey with a storage capacity of 31000 hm³.and 207 m height from foundation [30]. Recently scientists believe the fact that the over one percent of reservoirs resulted to earthquake which can damage or fail the main structure. It is not a negligible value that this mechanism should be considered by engineers in design stage.

Damages due to RTS have been in two dams: (1) Koyna dam, which is gravity dam having 103 m height in India. It was subjected to an earthquake with magnitude of 6.3 in 1967. (2) Hsinfengkiang dam, which a buttress dam having a height of 105 m in China. It was subjected to earthquake with magnitude of 6.1 in 1962. Researchers state that earthquakes were caused in their reservoirs by RTS. The substantial longitudial cracks were developed near crest for both dams. Both dams are still in operation after strengthened.

The reservoir capacity is an important factor in triggering earthquakes as well as reservoir depth. Phenomenon about Reservoir Induced Seismicity (RIS) mainly conforms for the reservoir filling periods. It can also be seen for a reservoir after a certain time lag [5].

There are some important cases that strong earthquakes may affect a large area. Recent surveys indicate that there are at least 100 cases of earthquakes, which were triggered by reservoirs. The most serious case may be the 7.9-magnitude Sichuan earthquake in May 12, 2008, which killed an estimated 90,000 people. This earthquake has been related to the construction of the Zipingpu Dam, which is a 156 m high concrete faced rockfill dam with a reservoir of 1 120 hm³. [7] classified two types of earthquakes associated with reservoirs while explaining the complicated mechanisms of RTS after the 12 May 2008 Wenchuan earthquake in China: (1) The small magnitude earthquakes, which occur immediately after reservoir impounding or following sudden reservoir water level fluctuations are mainly related to stress adjustments in the foundation rock, collapse of karst caves and mining pits and mass movements, (2) Earthquakes, which are caused by seismicgenic faults passing through or adjacent to the reservoir area, are referred to as RTS. [7] states that the initial stress state must already be close to failure so that a minor change in strength properties in a fault plane caused by water in the reservoir could trigger seismic events and the magnitude of RTS events may gradually increase until the main shock occurs. Authors have explained the mechanism of Wenchuan earthquake by the earthquake of tectonic nature.

A CASE STUDY ON THE RESERVOIR TRIGGERED SEISMICITY FOR AN EXISTING DAM

Turkey is one of the most seismically active regions in the world. There are so many dams, which are under the effect of near-source zones in Turkey. There are some examples of embankment dam in Turkey, which were damaged during the earthquakes occurred in past. There is no any concrete dam, which was damaged as a result of earthquake in Turkey [25, 26].

Ataturk dam, which is a 169 m height zoned rockfill dam on the Euphrates River in Turkey with an 84 000 hm^3 of water reservoir, poses high risk about triggering phenomena by reservoir. It has the largest reservoir of Turkey with 48 700 hm^3 (Figure 2). Its crest length is 1 670 m and base width is approximately 900 m. It is located 35 km north of the Birecik dam reservoir and 120 km south of Karakaya dam body.

Its main embankment construction was started in 1985 and completed in 1990. The reservoir level has maximally reached to 537 m up to now. Its level fluctuates from 526 to 535 m as based on climate change and energy demand. It was designed a multi-purpose structure for irrigating lands, producing electricity and providing flood control. It generates electricity of 8100 GWh per year with an installed capacity of 2400 MW.

Figure 2: A general view from Ataturk dam.

It is a rockfill dam with central core. There is a transition section of sand, gravel and small sized crushed rock between the core and rockfill materials. It has also a coarse grained soil zone obtained from river deposits and a random zone, which is composed of laminated limestone. The upstream and downstream shells are composed of large-sized crushed rocks. The alluvium on river bed, which is composed of sand, gravel, clay and silt mixtures, was removed before beginning the construction of the main embankment. The basement of Ataturk dam is formed by karstic limestone, regarded as problematic rock for dam foundation. An intensive grouting program was applied to prevent water leakage from the reservoir [28].

Author has completed a study from seismic hazard analysis to 2-D finite element analysis to assess its static and dynamic stability for Ataturk dam [31]. For the seismic hazard analyses of the dam site, first all possible seismic sources were identified as based on the new seismic zonation map of Turkey by means of a software, which was developed at the Earthquake Research Center in Eskisehir Osmangazi University [29]. As a result of detailed evaluation, the dam site and vicinity were separated into four seismic zones. Figure 3 shows these zones including faults and earthquakes occurred in the basin along last 100 years.

At the first design stage, it was considered only Eastern Anatolian Fault System (DAF) for seismic hazard analysis. As a result of this study, the value of Peak Ground Acceleration (PGA) was low for MDE. Recent study conducted by author indicates that the PGA value is considerable level and Bozova fault has a significant potential for reservoir triggering seismicity for Ataturk dam. It was located 3.0 m far away from the dam body and has a parallel position to the dam crest. This fault can produce an earthquake with a magnitude of 6.5 to 7.0. The seismic hazard analysis was performed for the dam by means of two separate methods. The deterministic seismic hazard analysis shows that the PGA value ranges from 0.284 to 0.536. These PGA values are high. Because the fault is very close to the dam site. The results of probabilistic seismic hazard analysis indicate that peak ground acceleration (PGA) changes within a wide range (0.057g and 0.203g) for OBE. For MDE and SEE, the PGA value averages to 0.197g and 0.408, respectively [31].

The seismic hazard analyses performed throughout this study indicates that Ataturk dam is one of the most critical dams within the basin. As based on the author's recent studies, Total Risk Factor (TRF) value is 146.5 and it is identified as risk class of III. It means that it has high risk potential for downstream life and structures. [31] states that the 25-years old rockfill dam also has some problems in static condition and it cannot meet current seismic design standards. The earthquake intensity in dam site and reservoir area has been increased after reservoir impounding or following sudden reservoir water level fluctuations. The Bozova fault, which is very close to dam body, can be a source of earthquake triggered by the reservoir of Ataturk dam. Also, Terbela dam with a reservoir of 13 690 hm^3 in Pakistan can be classified as high risk dam when considered this phenomena.

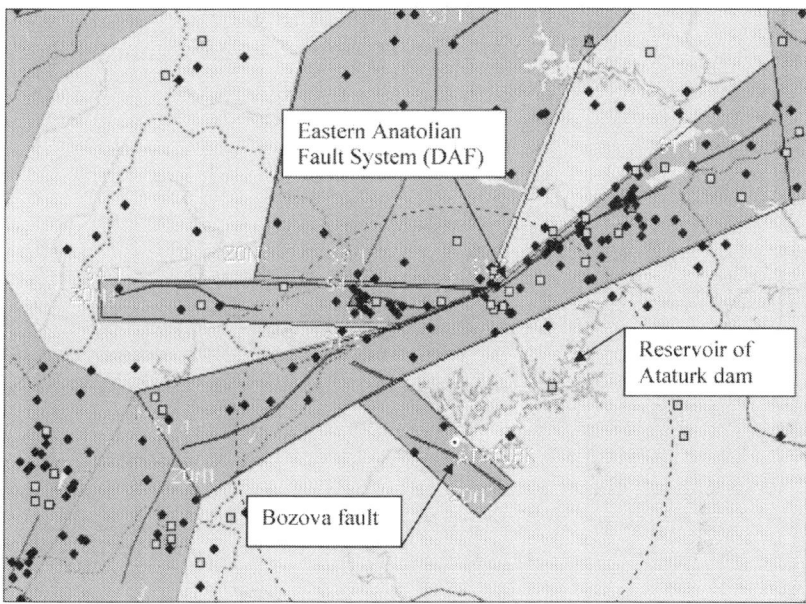

Figure 3: Location of dam site on seismo-tectonic map and earthquakes.

MAIN DESIGN PRINCIPLES FOR DAMS LOCATED ON ACTIVE SEISMIC AREA

The seismic activity of dam sites is generally analyzed by two methods: (1) The deterministic seismic hazard analysis and (2) The probabilistic seismic hazard analysis. The deterministic seismic hazard analysis is a very simple procedure and gives rational solutions for large dams. Due to the unavailability of strong motion records, various attenuation relationships were adopted to calculate the peak ground acceleration (PGA) acting on dam sites. The probabilistic seismic hazard analysis considers uncertainties in size, location and recurrence rate of earthquakes. The probabilistic seismic hazard analysis provides a framework in which uncertainties can be identified and combined in a rational manner to provide a more complete picture of the seismic hazard [11].

The computer program used for seismic analysis should be available for the probabilistic and deterministic assessment of seismic hazard. The seismic sources should be identified and the recurrence interval of earthquakes should be estimated. As a result of an extensive survey and a search of available literature [28, 29, 31]. Several sources have been identified to help analyzing the seismic hazard of dams in Turkey and surrounding countries. The earthquakes that occurred within the last 100 years should be used for estimating seismic parameters. Seismic zones and earthquakes within the area having a radius of 100 km around the dam site should be considered.

For beginning to a seismic hazard analysis, primary factors such as regional geological setting, seismic history and local geological setting should be studied, and then earthquake definitions should be executed. Figure 4 summarizes the methods of analysis for a dam site and dam body. After selection of earthquakes, static, pseudo-static and dynamic analyses should be performed and liquefaction phenomenon and near source effect should be evaluated. In Turkey, a design engineer should conform to diagram given in Figure 4.

Earthquake definitions are given below:

The Operating Basis Earthquake (OBE) was defined by means of the probabilistic methods. It is known as the earthquake that produces the ground motions at the site that can reasonably be expected to occur within the service life of the project [3]. It will be appropriate to choose a minimum return period of 145 years. It means a 50 percent probability of not being exceeded in 100 years.

Maximum Credible Earthquake (MCE), which is the largest earthquake magnitude that could occur along a recognized fault or within a particular seismo-tectonic province, was obtained for each zone and the most critical framework for the dam site was selected as Controlling Maximum Credible Earthquake (CMCE). The Maximum Design Earthquake (MDE) was then defined. It generally represents the ground motion with 475 years of return period [28]. It means a 10 percent probability of not being exceeded in 50 years.

According to [3], MDE is normally characterized by a level of motion equal to that expected at the dam site from the occurrence of deterministically evaluated MCE and Safety Evaluation Earthquake (SEE) should be used for critical structures located in very active seismic area. Most of large dams in Turkey were analyzed by using these definitions [28].

Terminology used for seismic analysis of dams varies between countries. In the last publication of [8], new earthquake definitions have been made. In this bulletin the Safety Evaluation Earthquake (SEE) is newly defined as the level of shaking for which damage can be accepted but for which there should be no uncontrolled release of water from the reservoir. In Turkey, it is defined as a level of ground motion having 2 percent probability of not being exceeded in 50 years. [8] states that SEE may be chosen to have a lower return period depending on the consequences of dam failure.

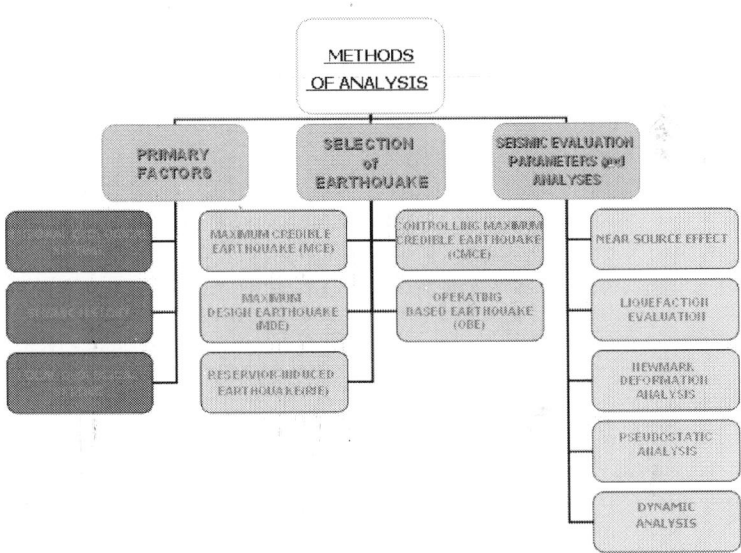

Figure 4: Methods of analysis for a dam located on active seismic area in Turkey

CONCLUSIONS

The total risk for dam structures mainly depends on the seismic hazard rating of dam site and the risk rating of the completed structure. This paper gives an overview on the effects of earthquake on dams and effects of dam on earthquake occurrence, and points out the necessity of special design and construction measures for the dams, which are under the effects of strong ground motions. It is clear that the main requirement in earthquake-resistant design for dams is to protect public safety, downstream life and property. Therefore, some important factors listed below should be taken into account in design stage:

i.   Large dams must be designed with a capability of resisting severe earthquake motion or fault movement at the dam site without uncontrolled release of water stored in reservoir.

ii.   For large dams located on non active seismic area, Reservoir Triggering Seismicity (RTS) can be more critical. Therefore, RTS should be defined sensitively as based on local geologic units and tectonic structures by means a detail seismic hazard analysis.

iii.   Damages to dams and their appurtenant facilities may be resulted from (1) direct fault movement across the dam foundation or (2) ground motion induced at the dam site by an earthquake located at some distance from the dam. The second one is commonly seen, however first one results to more serious problems for dams and their appurtenant structures.

iv.   Active faults pose the potential to cause damaging displacement of the structure when they are located very close to dam site. There are some examples of dams, which were damaged during the earthquakes that occurred in the past. Especially Concrete Faced Rockfill Dams (CFRD's) have high risk potential when considered near source effect with strong ground motion. This phenomenon is dealed with the fact that the transferred energy by rockfill is not absorbed by concrete face during earthquake. Near source effect should be considered with more attention for large dams in design stage.

v.   For the dams, which are under near source effect, embankment type with clay core seems more appropriate type because of being self-repairing properties of clay material when this type is technically and economically feasible for the selected dam site.

ACKNOWLEDGEMENTS

The author expresses his gratitude to the authorities of State Hydraulics Works for providing some technical data during completion of this study.

REFERENCES

1. Bell, M.L., Nur, A., 1978, Strength changes due to reservoir-induced pore pressure and stresses and application to Lake Oroville. J. Geophys. Res. 83, 4469– 4483.

2. Castro, G., Poulos, S.J., Leathers, F., 1985, Re-examination of slide of Lower San Fernando Dam. ASCE, Journal of Geotechnical Engineering 111. J. Geophys. Res. 83, 4469– 4483.

3. FEMA, 2005, Federal guidelines for dam safety-Earthquake analyses and Design of dams. Federal Emergeney Management Agency.

4. Gupta, H.K., 1992, Reservoir-Induced Earthquakes. Elsevier, Amsterdam, 364 pp.
5. Gupta, H.K., 2002, A review of recent studies of triggered earthquakes by artificial water reservoirs with special emphasis on earthquakes in Koyna, India. Earth-Science Reviews 58 (2002) 279–310
6. Houqun, C., 2008, Consideration of Dam Safety after Wenchuan Earthquake in China. The 14th World Conference on Earthquake Engineering October 12-17, 2008, Beijing, China.
7. Houqun, C., Zeping, X. and Ming., L, 2010, The relationship between large reservoirs and seismicity. International Water Power and Dam Construction, January, 29-32.
8. ICOLD, 2010, Selecting seismic parameters for large dams. Guidelines, Revision of Bulletin 72, Committee on Seismic Aspects of Dam Design, International Commission on large Dams, Paris,
9. Jansen, R.B.,(Ed.), 1988, Advanced Dam Engineering for Design, Construction and Rehabilitation. Van Nostrand Reinhold, New York. 884 pp.
10. JSDE, 2004, Preliminary Report of JSDE Investigation Team on the Niigata-Ken Cheetsu Earthquake. November.
11. Kramer, S.L., 1996, Geotechnical Earthquake Engineering. Prentice-Hall, Upper Saddle River, NJ. 653 pp.
12. Poulos, S.J. 1988, Liquefaction and related phenomena in Advanced Dam Engineering for Design, Construction and Rehabilitation (edited by Robert B. Jansen). Van Nostrand Reinhold, New York. 884 pp.
13. Rajendran, K., Harish, C.M., 2000, Mechanism of triggered seismicity at Koyna: an assessment based on relocated earthquake during 1983– 1993. Curr. Sci. 79 (3), 358– 363.
14. Rajendran, K. And Talwani, P., 1992, The role of elastic, undrained and drained responses in triggering earthquakes at Monticello Reservoir, South Carolina. Bull. Seismol. Soc. Am. 82, 1867–1888.
15. Roeloffs, E.A., 1988, Fault stability changes induced beneath a reservoir with cyclic variations in water level. J. Geophys. Res. 93 (B3), 2107– 2124.
16. Seed, H.B., Lee, K.L. and Idriss, I.M., 1969, Analysis of Sheffield Dam failure. Journal of the Soil Mechanics and Foundations Division, ASCE 95 (SM6), 1453–1490 (November).
17. Seed, H.B., Lee, K.L., Idriss, I.M. and Makdisi, F.I., 1975, The slides in the San Fernando Dams during the earthquake of February 9, 1971. Journal of the Geotechnical Engineering Division, ASCE 101(GT7), 651–688.
18. Scawthorn, C. 2002. Earthquake- A historical perspective in Earthquake Engineering Handbook (edited by Charles Scawthorn and Wai-Fah Chen).
19. Simpson, D.W., 1976, Seismicity changes associated with reservoir loading. Eng. Geol. 10, 123– 150.
20. Simpson, D.W., 1986, Triggered earthquakes. Annu. Rev. Earth Planet Sci. 14, 21– 42.
21. Simpson, D.W., Leith, W.S. and Scholz, C.H., 1988, Two types of reservoir-induced seismicity. Bull. Seismol. Soc. Am. 78 (6), 2025–2040.
22. Snow, D.T., 1982, Hydrology of induced seismicity and tectonism: case histories of Kariba and Koyna. Geol. Soc. Am., Spec. Pap. 189, 317–360.

23. Talwani, P., 1997, On the Nature of Reservoir-induced Seismicity. Pure Appl. Geophys. 150 (1997) 473–492.
24. Tosun, H., 2002, Earthquake-Resistant Design for Embankment Dams. Publication of General Directorate of State Hydraulic Works, Ankara. 208 pp. (in Turkish).
25. Tosun, H., Savaş, H., 2005. Seismic hazard analyses of concrete dams in Turkey. CDA Annual Conference on 100 years of Dam Experience— Balancing Tradition and Innovation, Calgary.
26. Tosun, H., Seyrek, E., 2005. Seismic hazard analyses and risk classification of large embankment dams in Turkey. Dam Safety 2005-Annual Conference of ASDSO, Orlando.
27. Tosun, H. and Seyrek, E. 2012. Selection of the appropriate methodology for earthquake safety assessment of dam structure, Advance in Geotechnical Earthquake Engineering-Soil Liquefaction and Seismic Safety of Dams and Monuments.
28. Tosun, H., Zorluer, İ., Orhan, A., Seyrek, E., Savaş, H. and Türköz, M. 2007a. Seismic hazard and total risk analyses for large dams in Euphrates Basin, Turkey. *Engineering Geology*, 89, 155-170.
29. Tosun, H., Türkoz, M. and Savas, H. 2007b, River basin risk analysis. Int. Water Power and Dam Construction, May issue.
30. Tosun, H., 2011, Earthquake safety evaluation of Keban dam, Turkey. CDA Annual Conference, Fredericton, Canada.
31. Tosun, H., 2012, Re-Analysis of Ataturk Dam under Ground Shaking By Finite Element Models. CDA Annual Conference, Saskatoon, Canada
32. Wieland, M., 2007, The seismic performance of concrete face rockfill dams under strong ground Shaking. International Water Power and Dam Construction, April, 18-20.
33. Wieland, M., 2013, Seismic design of major components. International Water Power and Dam Construction, February, 16-19.
34. Wieland, M., 2014, Seismic hazard and seismic design and safety aspects of large dam projects. Second European Conference on Earthquake Engineering and Seismology, Istanbul, Aug. 25-29, 2014.

CITATION

Hasan Tosun (2015). Earthquakes and Dams, Earthquake Engineering - From Engineering Seismology to Optimal Seismic Design of Engineering Structures, Prof. Abbas Moustafa (Ed.), ISBN: 978-953-51-2039-1, InTech, DOI: 10.5772/59372.

CHAPTER 4

Stability and Run-out Analysis of Earthquake-induced Landslides

Yingbin Zhang[1]

[1] Department of Geotechnical Engineering, School of Civil Engineering, Southwest Jiaotong University, Chengdu, China

INTRODUCTION

A large number of landslides can be caused by a strong earthquake and they have been the source of significant damage and loss of people and property. Therefore, it is very important to predict the stability of slope and the movement behaviors of a potential landslide under an earthquake loading, i.e., stability and run-out analysis (Figure 1).

Earthquake-induced landslides have been the source of significant damage and loss of people and property. One of the most serious event is the 1970 Peru earthquake. This event caused a huge rock avalanche that killed almost 54,000 people and buried two cities [143]. Another example is, in the 1920 Haiyuan earthquake, a large number of landslides caused widespread damage to infrastructure and buildings and killed at least 100,000 people, almost half of the total earthquake deaths [82].

Therefore, it is very important to predict the earthquake-induced landslides and to take countermeasures for potential landslides.

Main topics of earthquake-induced landslides are the following:

1. Investigation of recent and historical earthquake-induced landslides and their impacts so as to produce inventories of historical earthquake-induced landslides.

2. Prediction of potential earthquake-induced landslides, including (i) failure mechanism and stability analysis of seismic slopes, (ii) movement mechanism and behaviours of earthquake-induced landslides, and (iii) Instrumentation and monitoring technologies for potential earthquake-induced landslides or post-earthquake landslides.

3. Preventive countermeasures for earthquake-induced landslides, including (i) Stabilization and disaster mitigation of earthquake-related landslides, (ii) risk assessment and management of earthquake-related landslides, and (iii) hazard map and early warning system for earthquake-related landslides

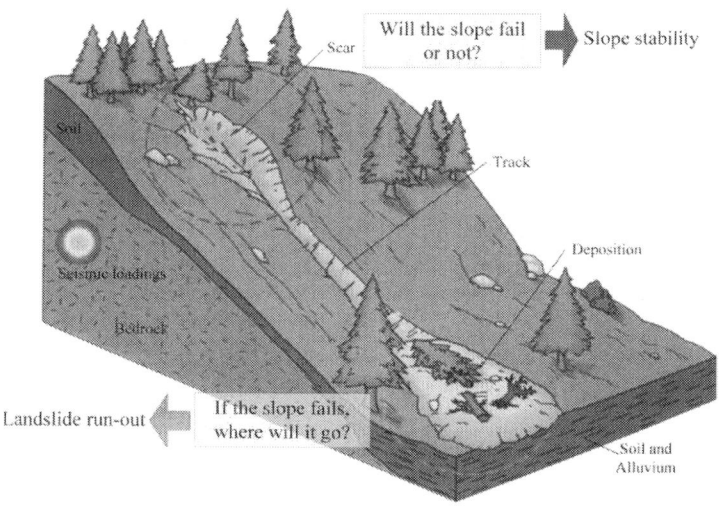

Figure 1: Stability and run-out analysis of earthquake-induced landslides.

This chapter focuses on the prediction of potential earthquake-induced landslides. The prediction of potential landslide can be carried out using detailed geotechnical investigations and stability calculations. (i) Failure mechanism and stability analysis of seismic slopes, i.e. seismic slope stability analysis and (ii) movement mechanism and behaviours of earthquake-induced landslides, i.e. landside run-out analysis are outlined firstly, and then the merits and demerits of each method are clarified in this chapter.

SEISMIC SLOPE STABILITY ANALYSIS

So far, methods developed to analyze the stability of earthquake slopes can be divided into three types: (1) pseudo-static methods, (2) dynamic sliding block methods, and (3) stress-strain methods. These three types of methods can be applied in different cases due to each of them has merit and demerit [73].

Pseudo-Static Methods

First presented the pseudo-static method, which is a simple method for evaluating of seismic stability of a slope. This type of method can be used to man-made or natural slopes based on either analytical method or numerical method. The earthquake force, acting on the an element or whole of the slope, is writed by a horizontal force and/or a vertical volum force equal to the gravitation force multiple a coefficient k, called the pseudo-static coefficient as shown in Figure 2 and Equation (1).

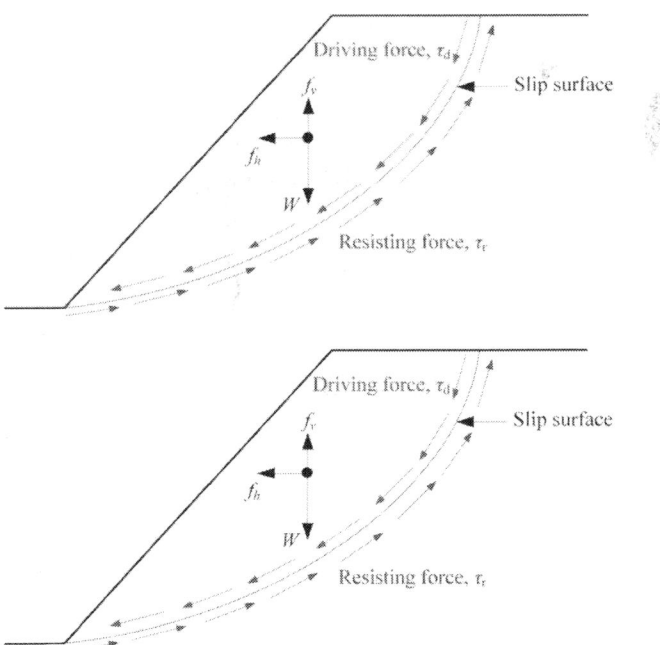

Figure 2: Forces acting on a slope in pseudo-static slope stability analysis.

Thus, k times the gravitational acceleration g, i.e. $a=kg$ forms the assumed seismic acceleration a. The assumed pseudo-static forces acting on a potential sliding mass of weight W will be

$$Fh = ah\ Wg = kh\ Wfv = av\ Wg = kvW \tag{1}$$

where ah and av are horizontal and vertical pseudo-static accelerations, respectively, kh and kv are horizontal and vertical pseudo-static coefficients, respectively. The factor of safety(FOS) is represented as the ratio of the resisting force to the driving force, Equation (2).

$$FOS = \tau r \tau d \tag{2}$$

From Equation (1), the pseudo-static force is determined by the seismic coefficient. The key problem for the pseudo-static procedure is how to select an appropriate seismic coefficient under an acceptable FOS. There have been studies for determining the most appropriate pseudo-static coefficient by a matter of experience and judgment.

[166] classical paper made the original suggestion to use of $kh=0.1$ for severe earthquakes, $kh=0.2$ for violent and/or destructive earthquakes, and of $kh=0.5$ for catastrophic earthquakes.

[103] presented a minimum pseudo-static FOS of 1.5 based on a slope material strength reduction factor (SRF) of 0.8 and the following acceleration values associated with two different earthquake magnitudes M. The same values of seismic coefficients for magnitude 6.5 and 8.25 earthquakes are recommended by [154], but with an acceptable FOS of 1.15.

$$A = 0.1g \text{ for } M = 6.5 \text{ implyingk} = 0.1a = 0.15g \text{ for } M = 8.25$$
$$\text{implyingk} = 0.15 \tag{3}$$

[137] also presented the pseudo-static coefficient related to earthquake magnitude. In detail, for an 8.25, 7.5, 7.0 and 6.5 magnitude earthquakes, if the seismic coefficients equal to 1/2, 1/3, 1/4 and 1/5 of the PGA, respectively, the computed FOSs are larger than 1.0, the accumulated displacements of slope are likely to be acceptably small.

In the report published by the International Commission of Large Dams (ICOLD), [154] shows a list of the minimum FOS value and horizontal seismic coefficients for 14 large dams worldwide, in which the minimum FOSs range from 1.0 to 1.5 and the horizontal earthquake coefficients range from 0.1 to 0.15. The Corps of Engineers Manual recommended a earthquake coefficient of 0.1 or 0.15 for areas where major and great earthquake threats are estimated, respectively, and a FOS of no larger than 1.0 for all magnitude earthquakes.

Some references related the earthquake coefficient value to the peak ground acceleration (PGA) [10,67, 108]. [108] related a pseudo-static coefficient of 1/3 to 1/2 of the PGA at the top of a double-side slope (a dam in the source reference), whereas [67] related a pseudo-static coefficient of 1/2 of the PGA of bedrock (PGA_{rock}) with a FOS of no larger than 1.0 and a SRF of 20%. And, [10] recommended the pseudo-static coefficient of 0.6 or 0.75 times of the PGA of bedrock (0.6 or $0.75PGA_{rock}$). It should be noted that the value given by [10] is conservative because the original study is designed for solid-waste landfills, where the allowable deformation are relatively small. [89] pointed that although engineering judgment is required for all cases, the criteria of [67] should be appropriate for most slopes.

[91] suggested one-half of PGA to use in an area of low seismicity (peak acceleration <0.15g) for the stability of earth embankments. This can be obtained from the peak horizontal motion (mean) from Modified Mercalli Intensity (MMI), magnitude-distance attenuation and the probability of a 50-year, 90% nonexceedance. However, in an area of moderate to strong seismicity ($0.15g \leq PGA \leq 0.40g$), PGA is obtained from the peak horizontal motion, from MMI, magnitude-distance attenuation and probability of 250-year, 90% nonexceedance.

[76] suggested a minimum FOS of 1.0, also based on a slope material SRF of 0.8 and the following values of pseudo-static coefficient: a equals to 0.17PGA or 0.5PGA for the dynamic response analysis is to be performed for the slope or earthquake structure or not.

[163] developed an expression for the earthquake coefficient in terms of characters of ground motion and magnitude of earthquake based on the data of [10].

It is almost common that only the horizontal acceleration is considered in evaluating the stability and deformation of a slope because the horizontal acceleration is the principal de-stabilizing force that acts on earth structures as well as the principal source of damage observed in earthquakes [4].

From Figure 2, the horizontal force clearly increases the driving force and decreases the FOS. The vertical pseudo-static force generally has less influence on the FOS than the horizontal pseudo-static force does because the vertical pseudo-static reduces both the driving and resisting forces. Hence, the effects of vertical seismic loading are frequently omitted in pseudo-static analysis [89].

Several investigators performed some analyses and have shown that the inclination of seismic loading have a significant influence on the seismic stability of slope by coupling the vertical and horizontal components of seismic force [20, 100].

In summary, pseudo-static method can be simply and directly used to identify the FOS and the critical seismic coefficient kc. In addition, performance of slope is closely related to permanent displacement, but the results of pseudo-static method are difficult to interpret the performance of slope after a seismic event because this method provides no information about permanent displacement. Because the pseudo-static analysis method provides only a rough assessment of seismic slope, it should be only used for the preliminary procedures. More accurate methods can be used to the followed process [73, 163,170].

Dynamic Sliding Block Methods

Displacement-based dynamic slidng block method is another alternative approach to evaluate the seismic slope stability, as permanent displacement is a useful index of slope performance, especially for those man-made slopes constructed for special purposes such as dams, embankments et al. This method has been widely used in earthquake geotechnical engineering.

In 1965, [119] proposed the dynamic sliding block method for estmating the permanent displacement of embankment affected by a seismic loading. In this method, sliding would be induced once the seismic loading exceed the critical seismic force of a potential failure surface as shown in Figure 3. The sliding would be accumulated until the end of seismic loading. We can evaluate the accumulated permanent displacement to assess the seismic stability of a slope.

Newmark's method showes that the yield acceleration of a potential block is a function of the FOS and slope angle, as:

$$Ac = (FOS-1)\, gsin\alpha \tag{4}$$

where ac is in terms of the gravity acceleration g; FOS is the static factor of safety; and α is the slope angle.

Since then, the method has been numerous extensions and applications. The section 2.2.1 and section 2.2.2 will give reviews for these two aspects, respectively. In addition, a regional scale application of the dynamic sliding block method is reviewed in section 2.2.3.

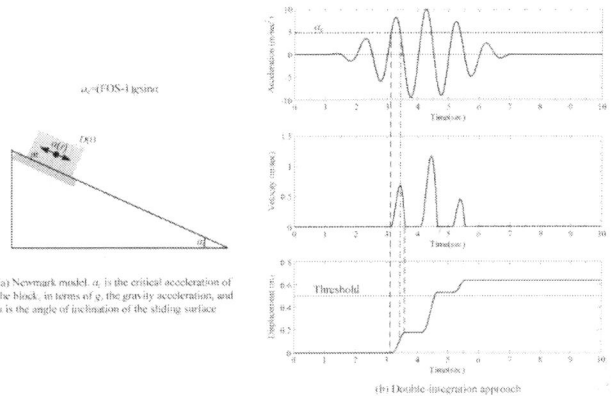

Figure 3: Illustration of the original Newmark's method.

Extensions

More attention has been focused over the last decades on developing methods to more accurately analyze the seismic stability of a slope for dams, embankments or other important structures by modeling the dynamic response of a slope more rigorously.

After the first dynamic sliding rigid block method, [155 and 97] published more sophisticated methods to account for the un-rigid block. Similar studies also given by [103]. As the classification given by [73], methods for estimating the permanent displacement of a sliding system induced by earthquake loading can been grouped into: (1) rigid-block model [119], (2) decoupled model [10, 104], and (3) coupled model [11, 97, 139].

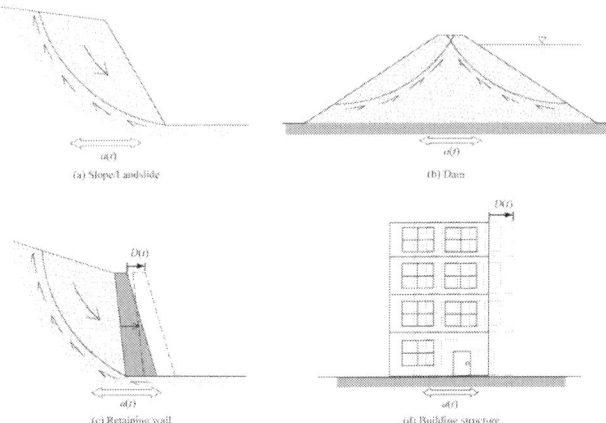

Figure 4: Applications of dynamic sliding block method in geotechnical engineering.

Applications

Since the rigid-block method was published in 1965 by Newmark, it has seen numerous applications, four of which are shown in Figure 4. The applications in recent years include (1) the seismic deformation analysis of earth dams and embankments [1, 22, 48, 49, 89, 90, 97, 103, 138, 144, 145, 150, 155, 179, 180]; (2) the displacements associated with landslides [34, 53, 70, 171]; (3) the seismic deformation of landfills with geosynthetic liners [10, 181]; (4) the seismic settlement of surface foundations [141]; and (5) the potential sliding of concrete gravity dams [32, 47, 95]. The extension of the analogue by [140] to gravity retaining walls has met worldwide acceptance, and has found its way into seismic codes of practice. Several other generalised applications have also appeared (e.g. [2, 3, 45,99, 139, 162, 169]).

Regional Scale Analysis

Except a single slope analysis, where the landslides are likely to occur and what kind of seismic conditions will cause it failure are two important topics in seismic hazard assessment, i.e.regional scale analysis [59].

For a regional scale analysis, slope stability analysis menthds will be not suitable [143, 168]. With the development of Geographic Information Systems (GIS) tools in recent years, regional scale analyses by the dynamic sliding block method have been proposed, in which ground shaking characteristic parameters, geotechnical material and topographic data are considered (e.g. [34, 71, 75, 106, 114, 151,155]).

The Newmark analysis (which combines slope stability calculations with seismic ground-motion records) is widely used to evaluate the potential for landslides that could be triggered by earthquake shaking [70, 71, 72, 74, 113].

Stress-Strain Methods

With the developments of the simulation approach and computer technology in recent years, the stress-strain method is becoming increasingly used in seismic slope stability analysis. These methods can be grouped into continuous methods, e.g. finite element method (FEM) [21], finite difference method (FDM) [116], boundary element method (BEM) [12], and discontinuous methods, e.g. rigid block spring method (RBSM) [77; 80], discontinuous deformation analysis (DDA) [159, 160] and discrete element method (DEM) [31].

Continuous Methods

[21] developed and named FEM of engineering analysis, in which the studied system is meshed into small many elements. This method can be applied to estimate the slope stability including dynamic stability analysis. Some applications of the continuous methods have been proposed, e.g. [89, 94, 153] and [156]. Recently, nonlinear in-elastic soil models have been developed and implemented in two-dimension (2-D) and three-dimension (3-D) models (e.g., [42, 50, 135, 164]). In addition, [93] and [183] studied the seismic slope stability by using FDM.

Discontinuous Methods

For the analysis of a potential failure mass consisting of multiple blocks as shown in Figure 5, the discontinuous methods are more applicable [120]. Some applications of RBSM and DEM can be found in some literature (e.g. [8, 52, 77, 79, 80, 85121, 127128, 129, 130, 131, 136, 182]).

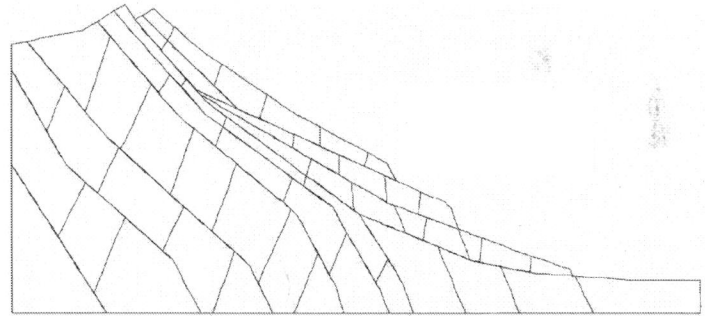

Figure 5: A jointed rock slope (modified from Bhasin and Kaynia, 2004)

DDA is also a discontinuous method developed for the modeling of the behaviors of multiple block systems. Since the novel formulation and the numerical code of DDA were presented, DDA draws more and more attention and many extensions and modifications to the original method have been proposed to overcome some limitations [19, 37, 38, 81, 87, 98] and make it more suitable, practical and efficient to seismic slope stability. The DDA can be used both to static rock slope engineering (e.g. [17, 81, 102, 123, 176, 187]) and the seismic rock slope stability analysis [56, 57, 54].

In summary, stress-strain method represents a powerful alternative approach for seismic slope stability analysis which is accurate, versatile and requires fewer a priori assumptions, especially, regarding the shape of failure surface.

LANDSLIDE RUN-OUT ANALYSIS

It is important to estimate the movement behaviour of a potential landslide. For example, the movement distance is an important parameter in risk assessment and measure design. There are many run-out analysis methods, which can fall into four categories: (1) experimental methods, (2) empirical methods, (3) analytical methods, and (4) numerical simulation methods. The states of the art of these methods are reviewed in the following four subsections 3.1 - 3.4.

Experiment Methods

Physical modelling typically involves using scale models to capture the motion of landslides. Physical experiments are usually preferred to models because models require more assumptions than direct measurements. But for landslides, direct experiment is difficult, dangerous, expensive, and of limited utility. Based on laboratory experiments and filed investigation data, there are many different available models developed for calculating run-out zones.

Some full-scale direct experiments with artificial landslides have been completed [118, 122124, 125,126] and others). However, since landslides are frequently heterogeneous and single event cannot be repeated carefully through adjusting only one factor, direct experiment is difficult, dangerous, expensive and of limited utility. And observing conditions are complicated by the danger of being in close proximity to a landslide and the difficulty of measuring a material with properties that change when observed in-situ or when isolated for measurement. But laboratory experiments are still the first qualitative and quantitative observations on the obtained results became fundamental for a better understanding of movement runout behaviour.

Empirical Methods

Several empirical methods for assessing landslide travel distance and velocity have been developed based on historical data and on the analysis of the relationship between parameters characterizing both the landslide, e.g. the volume of the landslide mass, and the path, e.g. local morphology, and the distance travelled by the failure mass [65]. Regression model-based methods and geomorphology-based methods are two kinds of common methods.

Regression Model-Based Methods

The regression model-based methods are developed on an apparent inverse relationship between landslide volume and angle of reach (also called as fahrböschund by [58]). Several linear regression equations have been proposed [25, 96, 153]. Introduced by [58], the angle of reach is the inclination of the line connecting the crest of the source with the toe of the deposit, as measured along the approximate streamline of motion. The angle of reach is considered an index of the efficiency of energy dissipation, and so is inversely related to mobility. Similar correlations between volume and other simple mobility indices have been proposed [33, 60, 142]. Given estimated source location, volume and path direction, these methods provide estimates of the distal limit of motion [111].

Improved empirical model notable performing regressions on subsets with varying scopes were presented by [13, 25, 69] and others.

Regression model-based models play a valuable role in landslide run-out analysis due to the regression model-based methods are simple. But the regression model-based methods are difficult to apply in practice with a high degree of certainty. For example, the correlation coefficients for some of regression models are 0.7-0.8, while a value of larger than 0.95 generally indicates a strong correlation. And it is difficult in this method to take account of influences of the ground condition, the micro-topography, the degree of saturation of the landslide mass and et al. For this point, geomorphology-based method is another alternative approach to predict the run-out of landslide.

Geomorphology-Based Methods

Field work and photo interpretation are the main sources of the geomorphological analysis for determining the travel distance of landslides [65]. The outer margin of the landslide deposits give an appraisal of the maximum distances that landslides have been able to reach during the present landscape (Figure 6). Several authors have provided these studies (e.g. [23, 24, 26, 88]).

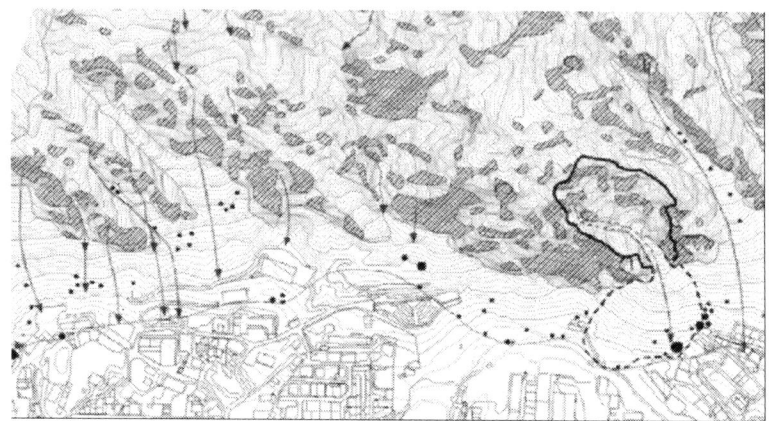

Figure 6: Boundary of the potential rockfall run-out area in Santa Coloma (Principality of Andorra), defined by the line that links the farthest fallen blocks observed in the field [24]. Arrows indicate historical rockfall paths and solid circles are large fallen boulders [65].

The geomorphological approach does not give any clue of the emplacement mechanism. Furthermore, the slope geometry and the circumstances responsible for past landslides might have changed. Therefore, results obtained in a given place cannot be easily exported to other localities.

In summary, empirical methods, both regression model-based methods and geomorphology-based methods, typically predict travel distances, while the deformation characteristics or the slide velocities of the landslide are not predicted. These models may be applied to establish initial hazard characteristics for preliminary run-out analysis, which may be later refined by other models.

Analytical Methods

In contrast to empirical methods, analytical methods are based on mechanics and involve the solution of motion equations [111]. The simplest analytical model is the classical sliding block model as shown in Figure 7, which is based on work-energy theory [6, 9, 43, Müller-Bernet in 58, 63, 83, 84, 132,147]. Internal deformation and it's associated energy dissipation are neglected and the landslides is treated as a lumped mass. At any position along the path, the sum of the energies including the potential energy, kinetic energy and net energy loss equals the initial potential energy. This energy balance can be visualized using the concept of energy grade lines, as shown in Figure 9. The concept of energy grade lines is useful for visualizing the energy balance. v is the velocity of the block, g is the vertical acceleration due to gravity and $v^2/2g$ is known as the velocity head, which is the kinetic energy of the block normalized by the product of its mass and g. The same normalization of net energy loss is known as head loss. Note that the positions of the energy lines are referenced to the centre of mass of the block and that the true energy line and mean energy line do not necessarily coincide. Given the initial position of the center of mass and a suitable relationship to approximate the energy losses, the position and velocity of the block can be determined at any given time.

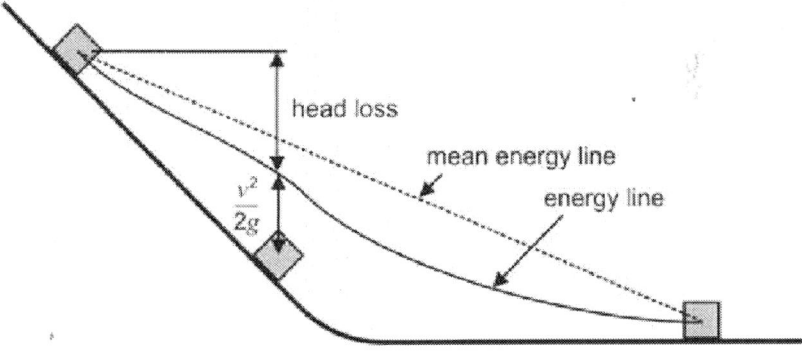

Figure 7: The classical sliding block model, based on work-energy theory [111]

Three-dimensional analysis for investigating runout of a slope were also proposed [36, 40, 51, 92 and109]. These models require a high resolution Digital Elevation Model (DEM).

Generally speaking, the use of analytical methods is somewhat motivated by the limitations of purely empirical methods, as the unique geometry and materials involved in each case can be accounted for explicitly and a statistically-significant database of previous events is not necessarily required. The simplicity of a lumped mass allows analytical solutions, fast and effectively [66]. However, because the landslide is reduced to a single point, lumped mass models cannot provide the exact maximum runout distance, but only the displacement concerning the centre of mass [44, 62].

Numerical Simulation Methods

The single-block model should be only applied to the motion of the center of mass of a rigid body, but more complex continuum deformable mass or multi-block system is often appeared in practice. Some numerical simulation methods have been developed to account explicitly for deformation during motion.

Continuous Methods

When considering that the dimensions of a typical particle is much smaller than the depth and length of the debris, the debris mass is treated as continuum. According to depth averaged Saint Venant approach, the material is assumed to be incompressible and the mass and momentum equations are written in a depth-averaged form. Many numerical methods now exist to investigate the run-out process of landslide (e.g. [18, 27, 28, 29, 30, 35, 62, 111, 112, 133, 149, 161, 165]). These methods are usually based on continuum mechanics and assume that the avalanche thickness is very much smaller than its extent parallel to the bed, i.e. thin layer depth-averaged models. The primary differences are their representation of basal resistance force and the constitutive relations describing the mechanical behaviour of the considered material. These models can accurately take account of detailed topography effects, shown to be significant, with a reasonable computational time, making it possible to perform sensitivity studies of the parameters used in the model. They can provide effective properties that make it possible to roughly reproduce not only the deposit shape but also the dynamic as shown in [46) and 117] for examples. However, conventional continuum approaching models, which neglects the contact between rocks, makes it impossible to trace the position of individual rock during a landslide.

Discontinuous Methods

When the landslide mass consists of large fragments and boulders, the run-out mass is modelled as an assembly of blocks moving down a surface. Some authors take circular shape models in their run-out analysis to evaluate maximum runout and final deposit position of past or potential events (e.g., [134]). Although polygonal shapes have the disadvantages due to the complexity of the contact patterns and penalty in computational time, methods using non-circular shapes will be required for more real-world problems. It is more appropriate when problems are limited in finite blocks. Discontinuous numerical simulation methods are powerful tools in simulation of failure and run-out process of rock avalanche controlled by weakness surface. DEM [31] and DDA [159, 157] are two of the most commonly used methods.

Both DEM and DDA employ the equations of dynamic motion which are solved at finite points in time, in a series of time steps, but there are some subtle but significant differences in their formulations of the solution schemes and contact mechanics. In the solution schemes, equations of motion in DDA are derived using the principle of minimization of the total potential energy of the system, while the equations of motion as implemented in DEM are derived directly from the force balance equations, which still resultant unbalanced force after a time step and damping is necessarily used to dissipate energy. In the contact mechanics, the DDA used a penalty method in which the contact is assumed to be rigid. No overlapping or interpenetration of the blocks is allowed as the same as real physical cases, whereas soft contact approach is used in DEM. The soft contact approach requires laboratory or field measured joint stiffness, which may be difficult to obtain in many cases. Many comparisons of basic models (sliding, colliding and rolling models) between the DEM and DDA were carried out and show that the results from DDA are more close to the analytical values than that from DEM [188]. Compared to DEM, DDA has a simpler and more straightforward physical meaning [172].

Applications of DEM can be found in some literatures, such as [85, 128, 129, 131, 136, 182].

DDA can be used for estimating the affected area of an earthquake-induced landslide [55] first validated the applicability of DDA for the dynamic behaviour of block sliding on an slope. Based on the same inputs model of seismic loadings, [7576886105, 158, 167, 178] studied the dynamic response or/and stability analysis of tunnel, slope, dam, foundation or ancient masonry structure by using DDA. Alternatively, the seismic loadings also can be applied to the base block [146, 147], which is different from the original DDA. Later, [173, 174, 175, 177] applied DDA to simulate the kinematic behavior of sliding rock blocks in the Tsaoling landslide and the Chiu-fen-erh-shan landslide induced by the 1999 Chi-Chi earthquake. Recently, [184] applied newest DDA program to simulate the largest landslides induced by the 2008 Wenchuan earthquake.

COMPARISONS OF VARIOUS METHODS

The studies in the field of the earthquake-induced landslides are generally reviewed. Two parts of contents, i) seismic stability analysis and ii) run-out analysis are reviewed and compared. Some conclusions can be drawn:

1. Three categories methods can be used to analyse the seismic stability of a slope. Each of these types of methods has strengths and

weaknesses and each can be appropriately applied in different situations. In detail, pseudo-static methods can simply and directly determine the FOS and the critical coefficient kc of a slope, while the widely used Newmark's methods and its extensions can determine the co-seismic deformation of a slope. And the Newmark's methods can be used to estimate where earthquake-induced landslides are likely to occur and what kind of shaking conditions will trigger them based on the GIS technology. More sophisticated analysis for real dynamic process of a seismic slope should be carried out by stress-strain methods, including both continuous methods and discontinuous methods.

2. Four kinds of methods can be used to analyse the run-out of a landslide. In detail, experiment method can provide the qualitative and quantitative observations on the obtained results although this method is difficult, dangerous, expensive and of limited utility. Empirical method can be directly used for assessing landslide travel distance and velocity based on historical data and on the analysis of the relationship between parameters characterizing both landslide and the path. Analytical method can be more directly used without the need of statistically-significant database of previous events. Numerical simulation method can be used to provide more information for the landslide composed by the complex continuum deformable mass or multi-blocks.

A CASE STUDY: THE DAGUANGBAO LANDSLIDE [185, 186]

Background

The Daguangbao landslide is located in the hanging wall only 6.5 km away from the Yingxiu-Beichuan fault. It is a typical bedding landslide. Figure 8 gives a pre- and post-earthquake 3-D topographies, from which cross-section of the Daguangbao landslide can be obtained (Figure 9). The extent of the damage caused by the Daguangbao landslide is reflected in the following statistics [61]:

1. The affected area covered 7.3~10 km²;
2. The accumulation body width is 2.2 km;
3. Estimated volume of collapsed rock mass is 750~840 million m³;

4.   The failure zone is more than 1 km;

5.   The failure mass moved about 4.5 km;

6.   Formed a 600m high landslide dam.

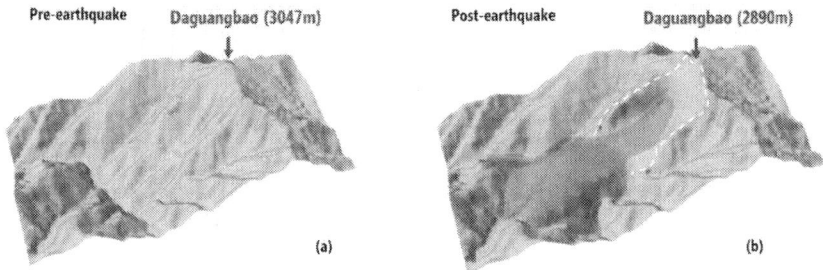

Figure 8: Pre- and post-earthquake 3-D topographies of the Daguangbao landslide.

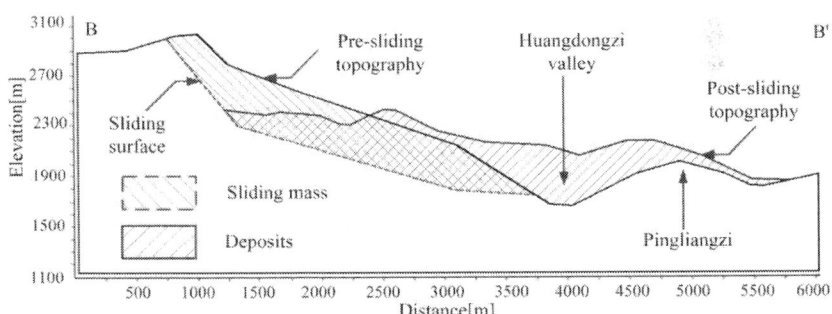

Figure 9: Cross-section of the Daguangbao landslide before and after the Wenchuan earthquake.

Material Properties And Ground Motion

Material Properties

The Daguangbao landslide is so huge that the size effect must be considered. To account for this discrepancy, experience equations based on Hoek-Brown failure criterion, which size effect can be considered, is used to back calculate the material strength. Table 1 lists the material properties of the Daguangbao landslide.

Table 1: Material properties of the Daguangbao landslide in FLAC^{3D} and DDA

	Materia l 1	Materia l 2	Materia 3 FLA C^{3D}	DDA
Density (ρ): g/cm^3	2.5	2.6	2.6	260000
Unit weight of rock (γ) : kN/m^3	25	26	26	0
Elastic modulus (E) : Gpa	1.86	2.63	14.76	
Poisson's ratio (v)	0.2	0.2	0.1	
Friction angle of discontinuities (φ) : $^\circ$	10.8	12.18	23.53	
Cohesion of discontinuities (c) : Mpa	1.276	1.576	4.052	
Tensile strength of discontinuities (σ_t) : kPa	12	32	556	

Ground Motion

The horizontal earthquake wave is the projection combination in the main sliding direction (N60°E) use the MZQP acceleration records in E-W and N-S directions as Equation (5). The inputted vertical earthquake wave is the MZQP acceleration records in U-D direction.

$$A\,H = aE - W \cdot \sin 60° + aN - S \cdot \cos 60° \qquad (5)$$

Figure 10 shows the input combined acceleration records. Velocity and displacement time histories can be obtained by first and second integration from acceleration record. The duration of earthquake wave is 60s.

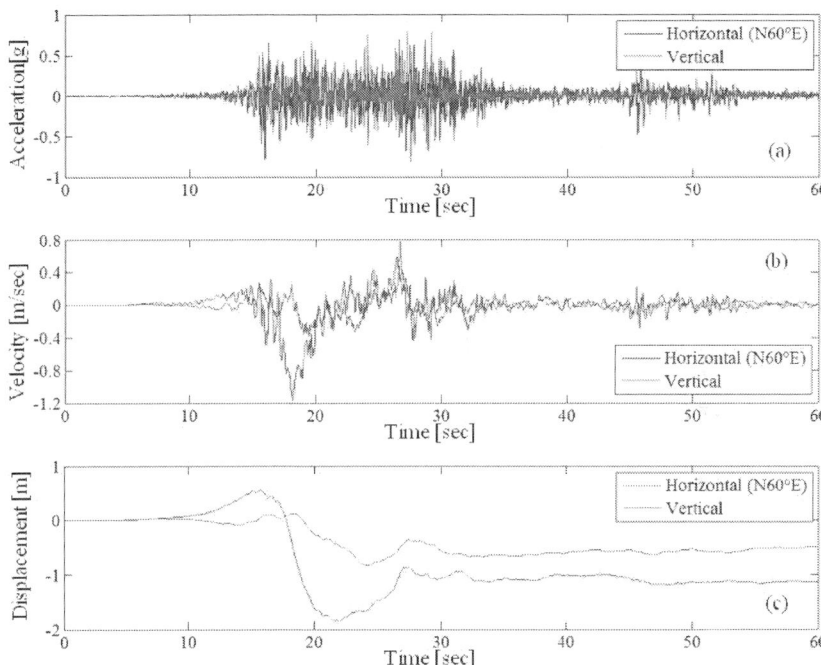

Figure 10: Input post-corrected horizontal and vertical ground records projected to N60°E direction.

Numerical Simulations-Run-Out Analysis

Seismic DDA can successfully simulate the movement of earthquake induced landslide. Two main features determine the Daguangbao landslide is a unique case, one is near-fault location (≈6.5 km) and the other one is huge scale (≈800×10^6 m^3). The near-fault location determines the Daguangbao landslide must be shocked by the extreme ground motion from the strong Wenchuan earthquake. And the Daguangbao landslide located on the meizoseismal area where the vertical seismic component is very large. In addition, the landslide is so huge that the size effects must be considered. The friction coefficient measured in the laboratory is no longer suitable for stability and run-out analysis.

To these two features, the Daguangbao landslide is simulated by the newest seismic DDA code in which multi-direction seismic forces can be applied in the base block directly, and experience equations based on Hoek-Brown failure criterion is applied to back-calculate the material strength by trying to consider the size effect.

Table 2: Control parameters for DDA

Parameter	Value
Assumed maximum displacement ratio (g_2)	0.001
Total number of time steps	20,000
Time step (g_1)	0.005s
Contact spring stiffness (g_0)	5.0×10^8 kN/m
Factor of over-relaxation	1.3

Geometry Of Sliding Blocks

The main sliding direction of the Daguangbao landslide, N60°E, is selected as analysis profile. The DDA model is depicted in Figure 11. In this simulation, based on the shape of failure surface and the character of slope topography, the whole slope is divided into three parts: base block, upper sliding mass, and lower sliding mass. Then two sliding masses are divided into the smaller discrete deformable blocks based on pre-existing discontinuities.

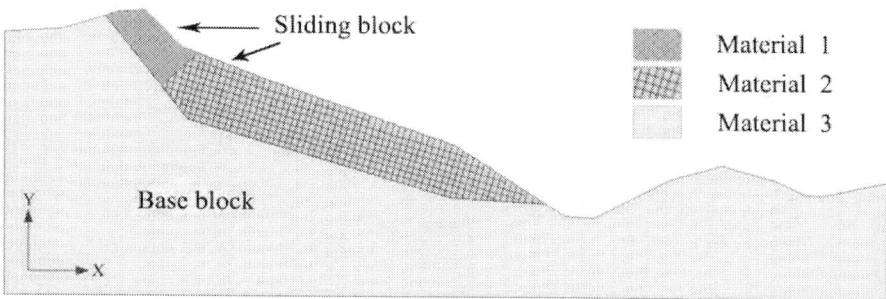

Figure 11: DDA model of the Daguangbao landslide.

Results

Figure 15 shows the post-failure behavior of the Daguangbao landslide simulated by the seismic DDA code. Simulated results show that the sliding blocks climb over the Pingliangzi. After overlapping the final step of DDA calculation with the topographic cross-section at the Daguangbao landslide, the deposit pattern of the simulated Daguangbao landslide under horizontal-and-vertical situation coincides well with local topography.

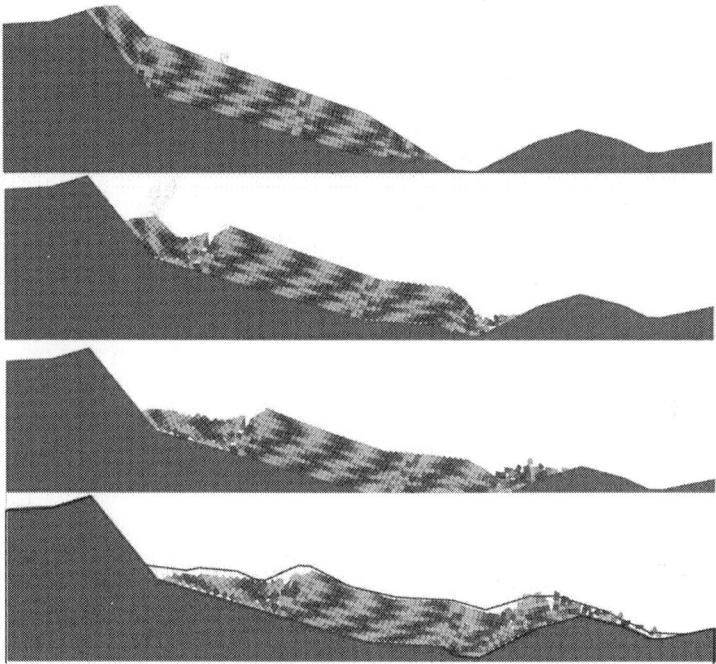

Figure 12: Simulation results of the Daguangbao landslide.

CONCLUSIONS

Five cases are performed using finite difference program FLAC^{3D}, under the real seismic waves near the study site. The results show that the seismic conditions cause a significant reduction in factor of safety than static situation. It also found that the vertical seismic has a significant influence on tension failure of block, although it has an insignificant influence on change of the factor of safety. Another important conclusion is the effect of vertical seismic force on relative displacement of potential sliding mass is significant. In addition, large area of tension failure caused by the combined seismic forces at back edge of the slope applies the evidence of effect of vertical seismic force on failure mechanism of slope stability.

A comparison of simulation results from three situations, static, only-horizontal and horizontal-and-vertical, is carried out. Seismic force has a significant influence on the arrival distance, and shape of post-failure. Arrival distance from horizontal-and-vertical situation is larger than that from only-horizontal situation. In addition, the deposit pattern of the simulated Daguangbao landslide under horizontal-and-vertical situation coincides well with local topography. The vertical seismic force should be considered for landslide assessment and management, especially in the situation that the studied site located on the meizoseismal area during the earthquake.

ACKNOWLEDGEMENTS

The authors gratefully acknowledge the financial support provided by the National Natural Science Foundation of China (No. 51408511), the Scientific Research Foundation for the Returned Overseas Chinese Scholars, State Education Ministry and the opening fund of State Key Laboratory of Geohazard Prevention and Geoenvironment Protection (Chengdu University of Technology) (No. SKLGP2014K015).

REFERENCES

1. Ambraseys, N., and Sarma, S. 1967. The response of earth dams to strong earthquakes. Geotechnique, 17(3): 181-213.
2. Ambraseys, N., and Menu, J. 1988. Earthquake-induced ground displacements. Earthquake Engineering & Structural Dynamics, 16(7): 985-1006.
3. Ambraseys, N., and Srbulov, M. 1994. Attenuation of earthquake-induced ground displacements. Earthquake Engineering & Structural Dynamics, 23(5): 467-487.
4. Anderson, D., and Kavazanjian Jr, E. 1995. Performance of landfills under seismic loading. In Proc., 3rd Int. Conf. on Recent Advances in Geotechnical Earthquake Engineering and Soil Dynamics. Univ. of Missouri Rolla, MO, Vol.3, pp. 277-306.
5. Army, U.C.o.E. 1960. Stability of Earth and Rockfill Dams, EM 1110-2-1902.
6. Azzoni, A., La Barbera, G., and Zaninetti, A. 1995. Analysis and prediction of rockfalls using a mathematical model. In International Journal of Rock Mechanics and Mining Sciences & Geomechanics Abstracts. Elsevier, Vol.32, pp. 709-724.
7. Bakun-Mazor, D., Hatzor, Y., and Glaser, S. 2012. Dynamic sliding of tetrahedral wedge: The role of interface friction. International Journal for Numerical and Analytical Methods in Geomechanics, 36(3): 327-343.

8. Bhasin R, Kaynia AM. 2004. Static and dynamic simulation of a 700-m high rock slope in western Norway. Engineering Geology, 71(3-4): 213-226.

9. Bozzolo, D., and Pamini, R. 1986. Simulation of rock falls down a valley side. Acta Mechanica, 63(1-4): 113-130.

10. Bray, J.D., and Rathje, E.M. 1998. Earthquake-induced displacements of solid-waste landfills. Journal of Geotechnical and Geoenvironmental Engineering, 124(3): 242-253.

11. Bray, J.D., and Travasarou, T. 2007. Simplified procedure for estimating earthquake-induced deviatoric slope displacements. Journal of Geotechnical and Geoenvironmental Engineering, 133(4): 381-392.

12. Brebbia, C.A., and Wrobel, L. 1980. The boundary element method. Computer methods in fluids.(A 81-28303 11-34) London, Pentech Press, Ltd., 1980: 26-48.

13. Cannon, S.H. 1993. An empirical model for the volume-change behavior of debris flows. In Hydraulic Engineering '93, San Francisco.

14. CDCDMG:California Department of Conversation, Division of Mines and Geology, 1997. Guidelines for evaluating and mitigating seismic hazards in California. CDMG Special Publication.

15. Chang, K.-T., Lin, M.-L., Dong, J.-J., and Chien, C.-H. 2011. The Hungtsaiping landslides: from ancient to recent. Landslides, 9(2): 205-214.

16. Chen, L., and Zhao, W. 1979. Longling earthquake, 1976. Earthquake press, Beijing.

17. Chen, G., and Ohnishi, Y. 1999. Slope stability analysis using Discontinuous Deformation Analysis method. Rock Mecganics for Industry: 535-541.

18. Chen, H., and Lee, C. 2000. Numerical simulation of debris flows. Canadian Geotechnical Journal, 37(1): 146-160.

19. Cheng, Y.M. 1998. Advancements and improvement in discontinuous deformation analysis. Computers and Geotechnics, 22(2): 153-163.

20. Chopra, A.K. 1966. The importance of the vertical component of earthquake motions. Bulletin of the Seismological Society of America, 56(5): 1163-1175.

21. Clough, R.W. 1960. The finite element method in plane stress analysis. In 2nd Conference on Electronic Computation, Pittsburgh, PA.

22. Constantinou, M., and Gazetas, G. 1987. Probabilistic seismic sliding deformations of earth dams and slopes. In Probabilistic Mechanics and Structural Reliability (1984). ASCE, pp. 318-321.

23. Copons, R., and Vilaplana, J.M. 2008. Rockfall susceptibility zoning at a large scale: From geomorphological inventory to preliminary land use planning. Engineering Geology, 102(3): 142-151.

24. Copons, R., Vilaplana, J.M., Corominas, J., Altimir, J., and Amigó, J. 2004. Rockfall Risk Management in High-Density Urban Areas. The Andorran Experience. Landslide hazard and risk: 675-698.

25. Corominas, J. 1996. The angle of reach as a mobility index for small and large landslides. Canadian Geotechnical Journal, 33(2): 260-271.

26. Costa, J.E. 1984. Physical geomorphology of debris flows. In Developments and applications of geomorphology. Springer. pp. 268-317.

27. Crosta, G., and Agliardi, F. 2003a. A methodology for physically based rockfall hazard assessment. Natural Hazards and Earth System Science, 3(5): 407-422.

28. Crosta, G., and Frattini, P. 2003b. Distributed modelling of shallow landslides triggered by intense rainfall. Natural Hazards and Earth System Science, 3(1/2): 81-93.

29. Crosta, G.B., Frattini, P., and Fusi, N. 2007. Fragmentation in the Val Pola rock avalanche, Italian Alps. Journal of Geophysical Research: Earth Surface (2003–2012), 112(F1).

30. Crosta, G., Imposimato, S., Roddeman, D., Chiesa, S., and Moia, F. 2005. Small fast-moving flow-like landslides in volcanic deposits: the 2001 Las Colinas Landslide (El Salvador). Engineering Geology, 79(3): 185-214.

31. Cundall, P. 1971. A computer model for simulating progressive, large scale movements in blocky rock system. In In Symposium of International Society of Rock Mechanics, Nancy, France, pp. 11–18.

32. Danay, A., and Adeghe, L. 1993. Seismic-induced slip of concrete gravity dams. Journal of Structural Engineering, 119(1): 108-129.

33. Davies, T.R. 1982. Spreading of rock avalanche debris by mechanical fluidization. Rock Mechanics, 15(1): 9-24.

34. Del Gaudio, V., Pierri, P., and Wasowski, J. 2003. An approach to time-probabilistic evaluation of seismically induced landslide hazard. Bulletin of the Seismological Society of America, 93(2): 557-569.

35. Denlinger, R.P., and Iverson, R.M. 2001. Flow of variably fluidized granular masses across three-dimensional terrain: 2. Numerical predictions and experimental tests. Journal of Geophysical Research: Solid Earth (1978–2012), 106(B1): 553-566.

36. Descoeudres, F., and Zimmermann, T. 1987. Three-dimensional dynamic calculation of rockfalls. In 6th ISRM Congress.

37. Doolin, D.M. 2005. Unified displacement boundary constraint formulation for discontinuous deformation analysis (DDA). International Journal for Numerical and Analytical Methods in Geomechanics, 29(12): 1199-1207.

38. Doolin, D.M., and Sitar, N. 2004. Time Integration in Discontinuous Deformation Analysis. Journal of Engineering Mechanics, 130(3): 249-258.

39. Dorren, L.K. 2003. A review of rockfall mechanics and modelling approaches. Progress in Physical Geography, 27(1): 69-87.

40. Dorren, L., and Heuvelink, G.B. 2004. Effect of support size on the accuracy of a distributed rockfall model. International Journal of Geographical Information Science, 18(6): 595-609.

41. Dreyfus, D., Rathje, E.M., and Jibson, R.W. 2013. The Influence of Different Simplified Sliding-Block Models and Input Parameters on Regional Predictions of Seismic Landslides Triggered by the Northridge Earthquake. Engineering Geology.

42. Elgamal, A.-W.M., Scott, R.F., Succarieh, M.F., and Yan, L. 1990. La Villita dam response during five earthquakes including permanent deformation. Journal of Geotechnical Engineering, 116(10): 1443-1462.

43. Evans, S., and Hungr, O. 1993. The assessment of rockfall hazard at the base of talus slopes. Canadian Geotechnical Journal, 30(4): 620-636.

44. Evans, S., Hungr, O., and Enegren, E. 1994. The Avalanche Lake rock avalanche, Mackenzie mountains, northwest territories, Canada: description, dating, and dynamics. Canadian Geotechnical Journal, 31(5): 749-768.

45. Fardis, M.N. 2009. Seismic design, assessment and retrofitting of concrete buildings: based on EN-Eurocode 8. Springer.

46. Favreau, P., Mangeney, A., Lucas, A., Crosta, G., and Bouchut, F. 2010. Numerical modeling of landquakes. Geophysical Research Letters, 37(15): L15305.

47. Fenves, G., and Chopra, A.K. 1986. Simplified analysis for earthquake resistant design of concrete gravity dams. Earthquake Engineering Research Center, University of California.

48. Franklin, A.G., and Chang, F.K. 1977. Permanent displacements of earth embankments by Newmark sliding block analysis.

49. Gazetas, G., and Uddin, N. 1994. Permanent deformation on preexisting sliding surfaces in dams. Journal of Geotechnical Engineering, 120(11): 2041-2061.

50. Griffiths, D., and Prevost, J.H. 1988. Two-and three-dimensional dynamic finite element analyses of the Long Valley Dam. Geotechnique, 38(3): 367-388.

51. Guzzetti, F., Malamud, B.D., Turcotte, D.L., and Reichenbach, P. 2002. Power-law correlations of landslide areas in central Italy. Earth and Planetary Science Letters, 195(3): 169-183.

52. Hamajima, R., Kawai, T., Yamashita, K., and Kusabuka, M. 1985. Numerical analysis of cracked and jointed rock mass. In the 5th International Conference on Numerical Methods in Geomechanics, Nagoya, Japan, pp. 207-214.

53. Harp, E.L., and Jibson, R.W. 1995. Inventory of landslides triggered by the 1994 Northridge, California earthquake. US Geological Survey.

54. Hatzor, Y.H. 2003. Fully Dynamic Stability Analysis of Jointed Rock Slopes. In Proceedings of the 10th ISRM Congress, pp. 503-514.

55. Hatzor, Y.H., and Feintuch, A. 2001. The validity of dynamic block displacement prediction using DDA. International Journal of Rock Mechanics & Mining Sciences.

56. Hatzor, Y.H., Arzi, A.A., and Tsesarsky, M. 2002. Realistic dynamic analysis of jointed rock slopes using DDA. In 5th Int. Conf. on Analysis of Discontinuous Deformation - Stability of rock structures, Abingdon, Balkema, Rotterdam, The Netherlands, pp. 47–56.

57. Hatzor, Y., Arzi, A.A., Zaslavsky, Y., and Shapira, A. 2004. Dynamic stability analysis of jointed rock slopes using the DDA method: King Herod's Palace, Masada, Israel. International Journal of Rock Mechanics and Mining Sciences, 41(5): 813-832.

58. Heim, A. 1932. Bergsturz und Menschenleben, Zurich: Fretz and Wasmuth Verlag.

59. Hsieh, S.-Y., and Lee, C.-T. 2011. Empirical estimation of the Newmark displacement from the Arias intensity and critical acceleration. Engineering Geology, 122(1-2): 34-42.

60. Hsü, K.J. 1975. Catastrophic debris streams (sturzstroms) generated by rockfalls. Geological Society of America Bulletin, 86(1): 129-140.

61. Huang, R., Pei, X., Fan, X., Zhang, W., Li, S., and Li, B. 2012. The characteristics and failure mechanism of the largest landslide triggered by the Wenchuan earthquake, May 12, 2008, China. Landslides, 9(1): 131-142.

62. Hungr, O. 1995. A model for the runout analysis of rapid flow slides, debris flows, and avalanches. Canadian Geotechnical Journal, 32(4): 610-623.

63. Hungr, O., and Evans, S. 1988. Engineering evaluation of fragmental rockfall hazards. In Proceedings of the Fifth International Symposium on Landslides, Lausanne, AA Balkema, Rotterdam, Netherlands, pp. 685-690.

64. Hungr, O., and Evans, S. 2004. Entrainment of debris in rock avalanches: An analysis of a long run-out mechanism. Geological Society of America Bulletin, 116(9-10): 1240-1252.

65. Hungr, O., Corominas, J., and Eberhardt, E. 2004. Estimating landslide motion mechanism, travel distance and velocity.

66. Hürlimann, M., Rickenmann, D., Medina, V., and Bateman, A. 2008. Evaluation of approaches to calculate debris-flow parameters for hazard assessment. Engineering Geology, 102(3): 152-163.

67. Hynes-Griffin, M.E., and Franklin, A.G. 1984. Rationalizing the seismic coefficient method. Defense Technical Information Center.

68. Ishikawa, T., Sekine, E., and Ohnishi, Y. 2002. Shaking table tests of coarse granular materials with discontinuous analysis. Proc. of ICADD-5, BALKEMA: 181-187.

69. Jakob, M., and Hungr, O. 2005. Debris-flow hazards and related phenomena. Springer.

70. Jibson, R.W. 1993. Predicting earthquake-induced landslide displacements using Newmark's sliding block analysis. Transportation Research Record: 9-9.

71. Jibson, R. 2000. A method for producing digital probabilistic seismic landslide hazard maps. Engineering Geology.

72. Jibson, R.W. 2007. Regression models for estimating coseismic landslide displacement. Engineering Geology, 91(2-4): 209-218.

73. Jibson, R.W. 2011. Methods for assessing the stability of slopes during earthquakes—A retrospective. Engineering Geology, 122(1-2): 43-50.

74. Jibson, R.W., and Jibson, M.W. 2003. Java programs for using Newmark's method and simplified decoupled analysis to model slope performance during earthquakes. US Department of the Interior, US Geological Survey.

75. Jibson, R.W., Harp, E.L., and Michael, J.A. 1998. A Method for Producing Digital Probabilistic Seismic Landslide Hazard Maps: An Example from the Los Angeles, California, Area.

76. Kavazanjian, E., and Consultants, G. 1997. Design Guidance: Geotechnical Earthquake Engineering for Highways. Design Principles. Federal Highway Administration.

77. Kawai, T. 1977. A new discrete analysis of nonlinear solid mechanics problems involving stability, plasticity and crack. In the Symposium on Applications of Computer Methods in Engineering, Los Angeles, USA, pp. 1029-1038.

78. Kawai, T. 1978. New discrete models and their application to seismic response analysis of structures. Nuclear Engineering and Design, 48(1): 207-229.

79. Kawai, T., Takeuchi, N., and Kumeta, T. 1981. New discrete models and their application to rock mechanics. In ISRM International Symposium.

80. Kawai, T., Kawabata, Y., Kumagai, K., and Kondou, K. 1978. A new discrete model for analysis of solid mechanics problems. Numerical methods in fracture mechanics: 26-37.

81. Ke, T.C. 1996. The issues of rigid-body rotation in DDA. In First international forum on discontinuous deformation analysis (DDA) and simulations of discontinuous media, Berkeley, USA, pp. 318-325.

82. Keefer, D.K. 2000. Statistical analysis of an earthquake-induced landslide distribution—the 1989 Loma Prieta, California event. Engineering Geology, 58(3): 231-249.

83. Kirby, M.J., and Statham, I. 1975. Surface stone movement and screen formation. Journal of Geology, 83(3): 349-362.

84. Kobayashi, Y., Harp, E., and Kagawa, T. 1990. Simulation of rockfalls triggered by earthquakes. Rock Mechanics and Rock Engineering, 23(1): 1-20.

85. Komodromos, P., Papaloizou, L., and Polycarpou, P. 2008. Simulation of the response of ancient columns under harmonic and earthquake excitations. Engineering Structures, 30(8): 2154-2164.

86. Kong, X., and Liu, J. 2002. Dynamic failure numeric simulations of model concrete-faced rock-fill dam. . Soil Dynamics and Earthquake Engineering 22(9–12): 1131–1134.

87. Koo, C.Y., and Chern, J.C. 1998. Modification of the DDA method for rigid block problems. International Journal of Rock Mechanics & Mining Sciences, 35: 683-693.

88. Kostaschuk, R. 1987. Identification of debris flow hazard on alluvial fans in the Canadian Rocky Mountains. Debris flows/avalanches: process, recognition, and mitigation, 7: 115.

89. Kramer, S.L. 1996. Geotechnical earthquake engineering. Prentice-Hall Civil Engineering and Engineering Mechanics Series, Upper Saddle River, NJ: Prentice Hall, c1996, 1.

90. Kramer, S.L., and Smith, M.W. 1997. Modified Newmark model for seismic displacements of compliant slopes. Journal of Geotechnical and Geoenvironmental Engineering, 123(7): 635-644.

91. Krinitzsky, E.L. 1993. Fundamentals of earthquake-resistant construction. Wiley. com.

92. Lan, H., Derek Martin, C., and Lim, C. 2007. RockFall analyst: A GIS extension for three-dimensional and spatially distributed rockfall hazard modeling. Computers & Geosciences, 33(2): 262-279.

93. Latha, G.M., and Garaga, A. 2010. Seismic Stability Analysis of a Himalayan Rock Slope. Rock Mechanics and Rock Engineering, 43(6): 831-843.

94. Lee, K.L. 1974. Seismic permanent deformation in earth dams, University of California, Los Angeles, CA.

95. Leger, P., and Katsouli, M. 1989. Seismic stability of concrete gravity dams. Earthquake Engineering & Structural Dynamics, 18(6): 889-902.

96. Li, T. 1983. A mathematical model for predicting the extent of a major rockfall. Zeitschrift Fur Geomorphologie, 24: 473-482.

97. Lin, J.S., and Whitman, R.V. 1983. Decoupling approximation to the evaluation of earthquake-induced plastic slip in earth dams. Earthquake Engineering & Structural Dynamics, 11(5): 667-678.

98. Lin, C.T., Amadei, B., Jung, J., and Dwyer, J. 1996. Extensions of discontinuous deformation analysis for jointed rock masses. international Journal of Rock Mechanics and Mining Sciences & Geomechanics Abstracts, 33(7): 671-694.

99. Ling, H.I. 2001. Recent applications of sliding block theory to geotechnical design. Soil Dynamics and Earthquake Engineering, 21: 189-197.

100. Ling, H., and Leshchinsky, D. 1998. Effects of vertical acceleration on seismic design of geosynthetic-reinforced soil structures. Geotechnique, 48(3): 347-373.

101. Lomnitz, C. 1970. Casualties and behavior of populations during earthquakes. Bulletin of the Seismological Society of America, 60(4): 1309-1313.

102. Luan, M., Li, Y., and Yang, Q. 2000. Discontinuous deformation computational mechanics model and its application in stability analysis of rock slope. Chinese Journal of Rock Mechanics and Engineering, 3: 006.

103. Makdisi, F.I., and Seed, H.B. 1977. Simplified procedure for estimating dam and embankment earthquake-induced deformations. In ASAE Publication No. 4-77. Proceedings of the National Symposium on Soil Erosion and Sediment by Water, Chicago, Illinois, December 12-13, 1977.

104. Makdisi, F.I., and Seed, H.B. 1978. Simplified procedure for estimating dam and embankment earthquake-induced failures. Journal of the Geotechnical Division, ASCE, 104: 849–861.

105. Makris, N., and Roussos, Y. 2000. Rocking response of rigid blocks under near-source ground motions. Geotechnique, 50(3): 243-262.

106. Mankelow, J.M., and MURPHY, W. 1998. Using GIS in the probabilistic assessment of earthquake triggered landslide hazards. Journal of Earthquake Engineering, 2(4): 593-623.

107. Marcuson, W. 1981. Moderator's report for session on Earth Dams and Stability of Slopes under Dynamic Loads. In Proceedings, International Conference on Recent Advances in Geotechnical Earthquake Engineering and Soil Dynamics, Vol.3, p. 1175.

108. Marcuson III, W.F., and Franklin, A.G. 1983. Seismic Design, Analysis, and Remedial Measures to Improve Stability of Existing Earth Dams, DTIC Document.

109. Masuya, H., Amanuma, K., Nishikawa, Y., and Tsuji, T. 2009. Basic rockfall simulation with consideration of vegetation and application to protection measure. Natural Hazards and Earth System Science, 9(6): 1835-1843.

110. McDougall, S. 2006. A new continuum dynamic model for the analysis of extremely rapid landslide motion across complex 3D terrain, University of British Columbia.

111. McDougall, D. 2006. The distributed criterion design. Journal of Behavioral Education, 15(4): 236-246.

112. McDougall, S., and Hungr, O. 2004. A model for the analysis of rapid landslide motion across three-dimensional terrain. Canadian Geotechnical Journal, 41(6): 1084-1097.

113. Meunier, P., Hovius, N., and Haines, A.J. 2007. Regional patterns of earthquake-triggered landslides and their relation to ground motion. Geophysical Research Letters, 34(20).

114. Miles, S.B., and Ho, C.L. 1999. Applications and issues of GIS as tool for civil engineering modeling. Journal of Computing in Civil Engineering, 13(3): 144-152.

115. Miles, S.B., and Keefer, D.K. 2000. Evaluation of seismic slope-performance models using a regional case study. Environmental & Engineering Geoscience, 6(1): 25-39.

116. Mitchell, A.R., and Griffiths, D.F. 1980. The finite difference method in partial differential equations(Book). Chichester, Sussex, England and New York, Wiley-Interscience, 1980. 281 p.

117. Moretti, L., Mangeney, A., Capdeville, Y., Stutzmann, E., Christian Huggel, C., Schneider, D., and Francois Bouchut, F. 2012. Numerical modeling of the Mount Steller landslide flow history and of the generated long period seismic waves. Geophys. Res. Lett., 39(L16402).

118. Moriwaki, H., Yazaki, S., and Oyagi, N. 1985. A gigantic debris avalanche and its dynamics at Mount Ontake caused by the Nagano-ken-seibu earthquake, 1984. In Proc. 4th Int. Conf. Field Workshop on Landslides, pp. 359-364.

119. Newmark, N.M. 1965. Effects of earthquakes on dams and embankments. Géotechnique, 15: 139-159.

120. Ning, Y., and Zhao, Z. 2012. A detailed investigation of block dynamic sliding by the discontinuous deformation analysis. International Journal for Numerical and Analytical Methods in Geomechanics: 1-21.

121. Niwa, K., Kawai, T., Ikeda, M., and Takeda, T. 1984. Application of a new discrete method to fracture analysis of brittle materials. In the 3rd International Conference on Numerical Methods in Fracture Mechanics, Swansea, U.K., pp. 13-27.

122. Ochiai, H., Okada, Y., Furuya, G., Okura, Y., Matsui, T., Sammori, T., Terajima, T., and Sassa, K. 2004. A fluidized landslide on a natural slope by artificial rainfall. Landslides, 1(3): 211-219.

123. Ohnishi, Y., Chen, G., and Miki, S. 1995. Recent development of DDA in rock mechanics. Proc. ICADD, 1: 26-47.

124. Okura, Y., Kitahara, H., and Sammori, T. 2000. Fluidization in dry landslides. Engineering Geology, 56(3): 347-360.

125. Okura, Y., Kitahara, H., Sammori, T., and Kawanami, A. 2000. The effects of rockfall volume on runout distance. Engineering Geology, 58(2): 109-124.

126. Okura, Y., Kitahara, H., Ochiai, H., Sammori, T., and Kawanami, A. 2002. Landslide fluidization process by flume experiments. Engineering Geology, 66(1): 65-78.

127. Pal, S., Kaynia, A.M., Bhasin, R.K., and Paul, D.K. 2011. Earthquake Stability Analysis of Rock Slopes: a Case Study. Rock Mechanics and Rock Engineering.

128. Papaloizou, L., and Komodromos, P. 2009. Planar investigation of the seismic response of ancient columns and colonnades with epistyles using a custom-made software. Soil Dynamics and Earthquake Engineering, 29(11-12): 1437-1454.

129. Papantonopoulos, C., Psycharis, I.N., Papastamatiou, D.Y., Lemos, J.V., and Mouzakis, H.P. 2002. Numerical prediction of the earthquake response of classical columns using the distinct element method. Earthquake Engineering & Structural Dynamics, 31(9): 1699-1717.

130. Pekau, O.A, and Yuzhu, C. 2004. Failure analysis of fractured dams during earthquakes by DEM. Engineering Structures, 26(10): 1483-1502.

131. Pekau, O.A, and Yuzhu, C. 2004. Seismic collapse behaviour of Damaged dams. In 13 WCEE: 13 th World Conference on Earthquake Engineering Conference Proceedings.

132. Pfeiffer, T.J., and Bowen, T. 1989. Computer simulation of rockfalls. Bulletin of the Association of Engineering Geologists, 26(1): 135-146.

133. Pirulli, M. 2005. Numerical modelling of landslide runout. A continuum mechanics approach, Politecnico di Torino.

134. Poisel, R., Preh, A., and Hungr, O. 2008. Run Out of Landslides–Continuum Mechanics versus Discontinuum Mechanics Models. Geomechanics and Tunnelling, 1(5): 358-366.

135. Prevost, J.H. 1981. DYNA-FLOW: a nonlinear transient finite element analysis program. Princeton University, Department of Civil Engineering, School of Engineering and Applied Science.

136. Psycharis, I., Lemos, J., Papastamatiou, D., Zambas, C., and Papantonopoulos, C. 2003. Numerical study of the seismic behaviour of a part of the Parthenon Pronaos. Earthquake Engineering & Structural Dynamics, 32(13): 2063-2084.

137. Pyke, R. 1991. Selection of Seismic Coefficients for Use in Pseudo-Static Slope Stability Analyses. http://www.tagasoft.com/Discussion/article2_html.

138. Rathje, E.M., and Bray, J.D. 1999. An examination of simplified earthquake-induced displacement procedures for earth structures. Canadian Geotechnical Journal, 36(1): 72-87.

139. Rathje, E.M., and Bray, J.D. 2000. Nonlinear coupled seismic sliding analysis of earth structures. Journal of Geotechnical and Geoenvironmental Engineering, 126(11): 1002-1014.

140. Richards, R., and Elms, D.G. 1979. Seismic behavior of gravity retaining walls. Journal of the Geotechnical Engineering Division, 105(4): 449-464.

141. Richards, J., Elms, D., and Budhu, M. 1993. Seismic bearing capacity and settlements of foundations. Journal of Geotechnical Engineering, 119(4): 662-674.

142. Rickenmann, D. 1999. Empirical relationships for debris flows. Natural Hazards, 19(1): 47-77.

143. Rodriguez, C.E., Bommer, J., and Chandler, R.J. 1999. Earthquake-induced landslides 1980-1997. soil Dynamics and Earthquake Engineering, 18: 325-346.

144. Sarma, S.K. 1975. Seismic stability of earth dams and embankments. Geotechnique, 25(4): 743-761.

145. Sarma, S.K. 1981. Seismic displacement analysis of earth dams. Journal of the Geotechnical Engineering Division, 107(12): 1735-1739.

146. Sasaki, T., Hagiwara, I., Sasaki, K., Ohnishi, Y., and Ito, H. 2007. Fundamental studies for dynamic response of simple block structures by DDA. In In Proceedings of the eighth international conference on analysis of discontinuous deformation: fundamentals and applications to mining & civil engineering, Beijing, China, pp. 141–146.

147. Sasaki, T., Hagiwara, I., Sasaki, K., Yoshinaka, R., Ohnishi, Y., and Nishiyama, S. 2004. Earthquake response analysis of rock-fall models by discontinuous deformation analysis. In In Proceedings of the ISRM international symposium 3rd ARMS, Kyoto, Japan, pp. 1267–1272.

148. Sassa, K. 1988. Motion of Landslides and Debris Flows: Prediction of Hazard Area: Report for Grant-in-aid for Scientific Research by Japanese Ministry on Education, Science and Culture (project No. 61480062). Disaster Prevention Research Institute.

149. Savage, S., and Hutter, K. 1989. The motion of a finite mass of granular material down a rough incline. Journal of Fluid Mechanics, 199(1): 177-215.

150. Sawada, T., Chen, W.F., and Nomachi, S.G. 1993. Assessment of seismic displacements of slopes. Soil Dynamics and Earthquake Engineering, 12: 357-362.

151. Saygili, G., and Rathje, E.M. 2009. Probabilistically based seismic landslide hazard maps: An application in Southern California. Engineering Geology, 109(3): 183-194.

152. Scheidegger, A.E. 1973. On the prediction of the reach and velocity of catastrophic landslides. Rock Mechanics, 5(4): 231-236.

153. Seed, H.B. 1973. Analysis of the Slides in the San Fernando Dams During the Earthquake of Feb. 9, 1971: Report to State of California Department of Water Resources, Los Angeles Department of Water and Power, National Science Foundation. College of Engineering, University of California.

154. Seed, H.B. 1979. Considerations in the earthquake-resistant design of earth and rockfill dams. Geotechnique, 29(3): 13-41.

155. Seed, H.B., and Martin, G.R. 1966. The seismic coefficient in earth dam design. Journal of Soil Mechanics & Foundations Div, 92(Proc. Paper 4824).

156. Serff, N. 1976. Earthquake induced deformations of earth dams. College of Engineering, University of California.

157. Shi, G.-H. 1988. Discontinuous Deformation Analysis A New Numerical Model for the Statics and Dynamics of Block Systems, University of California, Berkeley.

158. Shi, G. 2002. Single and multiple block limit equilibrium of key block method and discontinuous deformation analysis. In Proceedings of the 5th International Conference on Analysis of Discontinuous Deformation. Rotterdam: AA Balkema, pp. 3-43.

159. Shi, G.-H., and Goodman, R.E. 1985. Two dimensional discontinuous deformation analysis. International Journal for Numerical and Analytical Methods in Geomechanics, 9: 541-556.

160. Shi, G.-H., and Goodman, R.E. 1989. Generalization of two-dimensional discontinuous deformation analysis for forward modelling. International Journal for Numerical and Analyrucal Methods in Geomechanics, 13: 359-380.

161. Sousa, J., and Voight, B. 1991. Continuum simulation of flow failures. Geotechnique, 41(4): 515-538.

162. Stamatopoulos, C. 1996. Sliding system predicting large permanent co-seismic movements of slopes. Earthquake Engineering & Structural Dynamics, 25(10): 1075-1093.

163. Stewart, J.P., Blake, T.F., and Hollingsworth, R.A. 2003. A Screen Analysis Procedure for Seismic Slope Stability. Earthquake Spectra, 19(3): 697.

164. Taiebat, M., Kaynia, A.M., and Dafalias, Y.F. 2011. Application of an Anisotropic Constitutive Model for Structured Clay to Seismic Slope Stability. Journal of Geotechnical and Geoenvironmental Engineering, 137(5): 492.

165. Takahashi, T., Momiyama, A., Hirai, K., Hishinuma, F., and Akagi, H. 1992. Functional correlation of fetal and adult forms of glycine receptors with developmental changes in inhibitory synaptic receptor channels. Neuron, 9(6): 1155-1161.

166. Terzaghi, K. 1950. Theoretical Soil Mechanics.

167. Tsesarsky, M., Hatzor, Y., and Sitar, N. 2005. Dynamic displacement of a block on an inclined plane: analytical, experimental and DDA results. Rock Mechanics and Rock Engineering, 38(2): 153-167.

168. Varnes, D.J., Landslides, t.I.A.E.G.C.o., and Slopes, O.M.M.o. 1984. Landslide hazard zonation: a review of principles and practice.

169. Wartman, J., Asce, M., Bray, J.D., and Seed, R.B. 2003. Inclined Plane Studies of the Newmark Sliding Block Procedure. Journal of Geotechnical and Geoenvironmental Engineering, 129(8): 673-684.

170. Wasowski, J., Keefer, D.K., and Lee, C.-T. 2011. Toward the next generation of research on earthquake-induced landslides: Current issues and future challenges. Engineering Geology, 122(1-2): 1-8.

171. Wilson, R.C., and Keefer, D.K. 1983. Dynamic analysis of a slope failure from the 6 August 1979 Coyote Lake, California, earthquake. Bulletin of the Seismological Society of America, 73(3): 863-877.

172. Wu, J.-H. 2003. Numerical analysis of discontinuous rock masses using discontinuous deformation analysis, Kyoto University, Kyoto, Japan.

173. Wu, J.-H. 2010. Seismic landslide simulations in discontinuous deformation analysis. Computers and Geotechnics, 37(5): 594-601.

174. Wu, J.-H., and Chen, C.-H. 2011a. Application of DDA to simulate characteristics of the Tsaoling landslide. Computers and Geotechnics, 38(5): 741-750.

175. Wu, J.-H., and Tsai, P.-H. 2011b. New dynamic procedure for back-calculating the shear strength parameters of large landslides. Engineering Geology.

176. Wu, A., Ren, F., and Dong, X. 1997. A study on the numerical model of DDA and its preliminary application to rock engineering. Chinese Journal of Rock Mechanics and Engineering, 16(5): 411-417.

177. Wu, J., Lin, J., and Chen, C. 2009. Dynamic discrete analysis of an earthquake-induced large-scale landslide. International Journal of Rock Mechanics and Mining Sciences, 46(2): 397-407.

178. Yagoda-Biran, G., and Hatzor, Y.H. 2010. Constraining paleo PGA values by numerical analysis of overturned columns. Earthquake Engineering & Structural Dynamics, 39(4): 463-472.

179. Yegian, M.K. 1991a. Seismic risk analysis for earth dams. ASCE.

180. Yegian, M.K., Marciano, E.A., and Ghahraman, V.G. 1991b. Earthquake-induced permanent deformations: probabilistic approach. Journal of Geotechnical Engineering, 117(1): 35-50.

181. Yegian, M., Harb, J., and Kadakal, U. 1998. Dynamic response analysis procedure for landfills with geosynthetic liners. Journal of Geotechnical and Geoenvironmental Engineering, 124(10): 1027-1033.

182. Zhang, C., Pekau, O.A., Jin, F., and Wang, G. 1997. Application of distinct element method in dynamic analysis of high rock slopes and blocky structures. soil Dynamics and Earthquake Engineering, 16: 385-394.

183. Zhang, Y., Chen, G., Zheng, L., Wu, J., and Zhuang, X. 2012a. Effects of vertical seismic force on the initiation of the Daguangbao landslide induced by the Wenchuan earthquake. In The 8th Annual Conference of International Institute for Infrastructure, Renewal and Reconstruction, Kumamoto, Japan, pp. 530-539.

184. Zhang, Y., Chen, G., Zheng, L., and Li, Y. 2012b. Numerical analysis of the largest landslide induced by the Wenchuan earthquake, May 12, 2008 using DDA. In International Symposium on Earthquake-induced Landslides, Kiryu, Japan.

185. Zhang, Y., G. Chen, L. Zheng, Y. Li and J. Wu. 2013. Effects of near-Fault Seismic Loadings on Run-out of Large-Scale Landslide: A Case Study. Engineering Geology 166, 216-236.

186. Zhang, Y., G. Chen, L. Zheng, Y. Li, X. Zhuang. 2014. Effects of vertical seismic force on initiation of the Daguangbao landslide induced by the 2008 Wenchuan earthquake, Soil dynamics and earthquake engineering (In press).

187. Zhao, S.L., Salami, M.R., and Rahman, M.S. 1997. Discontinuous Deformation Analysis Simulation of Rock Slope Failure. In 9th International Conference on Computer Methods and Advances in Geomechanics, Wuhan, China

188. Zheng, L. 2010. Development of new models for landslide simulation based on discontinuous deformation analysis, Kyushu University, Fukuoka, Japan.

CITATION

Yingbin Zhang (2015). Stability and Run-out Analysis of Earthquake-induced Landslides, Earthquake Engineering - From Engineering Seismology to Optimal Seismic Design of Engineering Structures, Prof. Abbas Moustafa (Ed.), ISBN: 978-953-51-2039-1, InTech, DOI: 10.5772/59439.

CHAPTER 5

Variation of Altitude Observed on the Occasion of the Tohoku Earthquake (M = 9.0) Occurred on March 11, 2011

Pietro Milillo[1], Tommaso Maggipinto[2], Pier Francesco Biagi[2]

[1]School of Engineering, University of Basilicata, Potenza, Italy
[2]Department of Physics, University of Bari, Bari, Italy
E-mail: pietro.milillo@unibas.it

ABSTRACT

Since October 1, 2010, a GPS receiver is put into operation at Tokai (Japan) in an experiment on Neutrino Physics (T2K). A significant variation of the altitude was detected from the beginning of March 2011, so that it has made worthwhile to investigate the possibility that such variations could be correlated to the Tohoku earthquake. In order to investigate in details this possibility, we analyzed the GPS data collected during 2011 by GEONet the GPS Earth Observation Network (GEONET). GEONET is the GPS network of Japan and consists of 1240 permanent stations. Preliminary results of the analysis seemed to show ten days before the earthquake, some possible anomalous behaviors of the stations. These anomalous behaviors were particularly relevant for stations of the network near the epicentral area. While co-seismic and post-seismic variations are widely expected, the anomalies recorded about ten days before the earthquake could be seriously considered among short-term precursors of the earthquake. In order to confirm this possibility, more detailed studies have been performed. In particular, GEONET currently makes available only daily solutions of the stations coordinates. On the contrary, it is very important to improve the time resolution just to understand the features of the anomalies till the last hours before the Earthquake. For this reason, we have performed an analysis to evaluate the coordinates and movement on hourly basis so improving the time resolution.

KEYWORDS

Earthquake; GPS; Time Series

INTRODUCTION

A wide variety of natural phenomena are detrimental for the natural environment and for the anthropic structures and the human being. Earthquakes are among the most hazardous natural phenomena and, in order to mitigate their effects, the singling out of their precursors is a task of primary importance. Generally speaking, the earthquake precursors can be classified into three categories, depending on the time scale we are concerned with: "long-term" (of the order of a few hundred years), "medium-term" (of the order of hundreds to a few years) and "short-term" (of the order of a few months to a few days). We will focus on the short-term precursors.

A seismic precursor is an "anomalous" variation in some physical-chemical parameters that occurs before the seismic event and that is clearly linked to it. The most recent results show that seismic precursors can be divided into two different categories: ground precursors [1-4], and atmospheric precursors [5]. The first refer to anomalies in physical-chemical parameters of the ground such as resistivity, gas content and ionic content of deep waters, earth magnetic field, ground deformation etc.; the second refer to anomalies in physical-chemical parameters of the atmosphere, such as temperature, gas composition, density etc. Generally, the ground precursors are highlighted by means of on-ground measurements while atmospheric precursors are revealed by satellites [5,6].

In this paper, we present the possibility of investigating one of the main ground precursors, i.e. the crustal deformation. In particular, we study the height variation of the ground observed on the occasion of the Tohoku Earthquake (M = 9.0) occurred on March 11, 2011. The heights are evaluated by means of GPS technique and in this sense such ground precursors are monitored by means of satellite techniques. The height data refer to different geographical areas of Japan and the data analysis as revealed possible preseismic, coseismic and postseismic effects in the crustal altitude.

DATA DESCRIPTION

We used data of GPS Earth Observation Network (GEONET), a dense GPS receiver network composed by about one thousand GPS observation sites with 25 - 30 km average distance between stations. This network, operated by Geospatial Information Authority of Japan (GSI), has been used for crustal-deformation monitoring and geodetic control.

Each site is equipped with a GPS receiver, the communication device and a backup battery, which is stored in a stainless steel pillar. A pillar is five meters high, standing on a concrete cubic basement two meters large. The receivers are tuned to sample dual band carrier phase data and code data every 30 seconds. All the collected data are archived into a database in RINEX (Receiver Independent Exchange) format. GPS data network solution process starts soon after data downloading of the stations is complete (max latency of 15'). The processing generally provides Quick, Rapid and Final solutions according to different analysis strategies [7]. Quick analysis is routinely executed every three hours with 6-hour observation data and it used especially for monitoring crustal movement. Rapid analysis is performed every day with 24-hour data and it is used for screening in case of emergencies. Final analysis is executed two weeks after observation and it is the most accurate analysis in GEONET. Quick and Rapid analysis use the IGS Ultra Rapid Orbits; while the final analysis is executed with IGS precise orbits [8]. In this work we used precise solutions and RINEX observation files. As first step we have processed the data to achieve daily solutions of the coordinates for a period of four months centered around the epoch of occurrence of the Earthquake (11th March 2011) obtaining a daily preliminary analysis. This kind of products have been used for a daily preliminary analysis in order to emphasize general behavior in occasion of the 11 March 2011 M = 9.0 Tohoku Earthquake. RINEX observation files have been used applying the sliding window technique hourly analysis in order to bring out more accurate trends 15 days before and after the Earthquake. The daily network processing of the GEONET network is conducted by GSI, with Bernese GPS processing software, also RINEX observation files used for hourly analysis have been analyzed with BERNESE software [9,10].

DATA ANALYSIS

After the preliminary step in which we have analyzed the daily solution provided by GEONET, we have, as second step, unwrapped the data to perform their reprocessing in order to achieve hourly solutions and, therefore, improve the time resolution in order to understand if abrupt events or discontinuities have been occurred during the investigated period.

Starting from January 1st, 2011, the daily data related to the altitude of all GPS station have been examined 2 months before and after 11th March 2011 identifying different behaviors in GPS Network heights.

An example of daily data plot, related to one GEONET GPS station, is shown in Figure 1.

Hourly solutions have been analyzed with Bernese Software in order to confirm trend and improve standard deviation error. The hourly analysis has been performed an a sub-network that consists of 12 fiducial stations (see Table 1) not very far from the Tsukuba station (TSKB IGS) which acts as a reference station; so we have a total number of 13 stations. Sliding window parameters are: 12 hours window with 3 hours sliding time.

All Processing steps have been performed using the Bernese Processing Engine (BPE) [10,11]. Both Precise Point Positioning (PPP) and Double-Difference Processing (RNX2SNX) blocks have been used for this analysis. Bernese Input Parameters are shown in Table 2. PPP is a special case of zero-difference processing, in this particular case it can be considered a preparatory step to double-differencing processing. PPP relies on precise orbit and clock information for deriving precise site coordinates and a receiver clock correction independently for each analyzed station and is based on undifferenced code and/or phase observations.

PPP is a fast and efficient way to produce station coordinates but it should be underlined that it is not

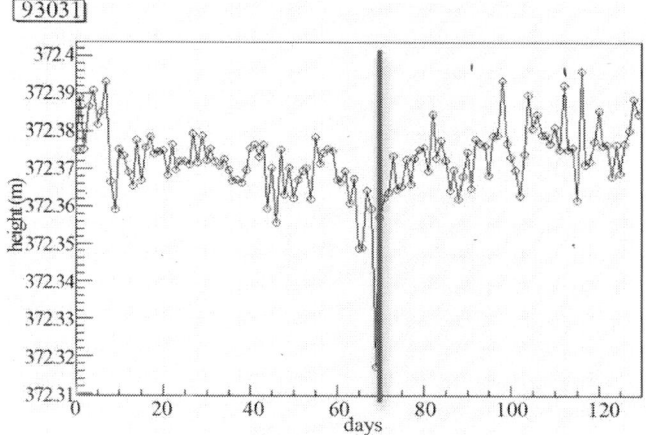

Figure 1. GPS-GSI daily data related to pre-seismic effect (Start day 1 January 2011).

Table 1. GPS network used for hourly analysis.

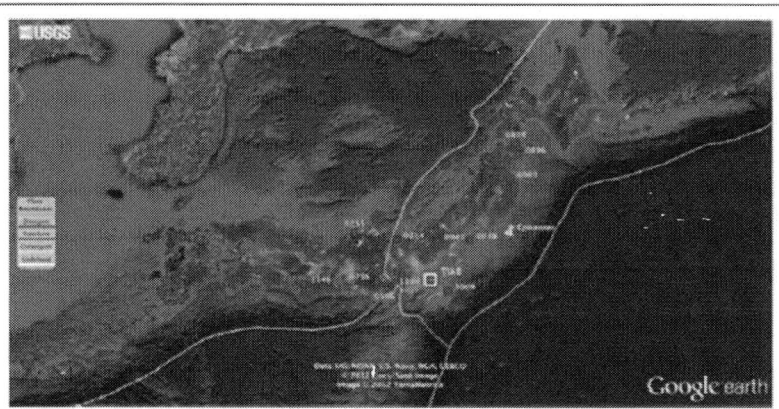

| 38 | 214 | 255 | 596 | 756 | 894 | 896 | 903 | 941 | 1100 | 1146 | 3009 | TSKB (IGS) |

Table 2. Bernese Input Parameters.

Bernese Input Parameters	
Precise Orbits	IGS
Tectonic Plate motion	NNR-NUVEL-1A
Solid Earth Tides, Pole Tides and Permanent Tides	IERS Conventions 2003
Ocean Tidal Loading Effects	Yes
Troposphere modeling estimation	Niell dry part

The PPP approach in our case is useful to identify problems and reject eventually GPS stations from the selected network before the RNX2SNX. The RNX2SNX is intended for a double-difference based analysis of RINEX GNSS observation data from a regional network. Station coordinates and troposphere parameters are estimated and stored in SINEX format to facilitate further processing and combination [10].

An important feature of this technique is that observation files with significant gaps or unexpectedly big residuals are automatically removed from the processing to ensure a robust processing and a reasonable network solution. The final network solution is a minimum constraint one, accomplished by three no-net translation conditions imposed on a set of ITRF2000 reference coordinates. The coordinates of all involved 12 fiducial stations are subsequently verified by means of a 7-parameter Helmert transformation in order to produce distortion-free transformations. In case of discrepancies, the network solution is recomputed based on a reduced set of fiducial stations. Height data as a function of time has been obtained after pre-processing step.

Non linear least square method with a function $x \cdot a + c$ (where a and c are computed parameters) has been computed on Daily and Hourly data in order to identify possible negative trend in occasion of the Seismic event.

RESULTS AND DISCUSSION

Daily Results

After plotting all 1400 GPS Network stations daily height data starting from January 1st, 2011, four different behaviors have been identified (Figure 2):

- Co-seismic effect
- Post-seismic effect
- Pre-seismic effect
- A So-called "January" effect Co-seismic effect is related to a large decrease of heights that occurs on the day of earthquake, while no other variation appears before or after the earthquake. This behavior has been revealed in many stations of the network generally located far enough from the epicentral area. These stations are reported in purple in the map (Figure 3).

Post-seismic effect concerns a permanent displacement that appears after the large decrease of altitude occurring the day of the earthquake. This behavior has been revealed in many stations of the network located mainly on the coastal direction near the epicentral area. Some of these stations are indicated in yellow in the map (Figure 3).

As for pre-seismic effect, about ten days prior the occurrence of the earthquake some suspect anomalous variation appears. This behavior was revealed in some stations of the network located mainly near the epicentral area.

Finally, during the first decades of January 2011 in many stations an evident variation of altitude (increase or decrease) appeared. In some case the variation was very large and this behavior was revealed mainly in stations located in the north of Japan. Red Squares in Figure 3 indicate GPS stations characterized by this kind of effect.

Hourly Results

Figure 4 shows the complete hourly sequence from the first of March to the end of March. As expected the result is comparable with the GSI elaborated daily data sequence. The only difference which characterizes GSI and ASI data is a negative constant of about 10 cm after the seismic event, this could depend from different parameters used for BERNESE software elaboration, GSI does not give its set of parameters but it seems that the constant shift depends on the choice of the reference point. In hourly data TSK has been set as the reference point and all GPS station positions have been calculated respect to this station. If, after the main shock, there is a TSK station shift this would appear on all GPS station measurements as a systematic error. We looked at TSK IGS elaborated data in order to confirm this hypothesis and what we found was a shift of exactly 10 cm (Figure 5).

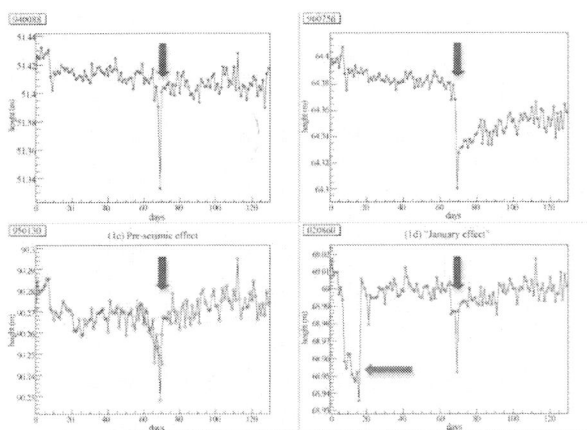

Figure 2. GPS-GSI daily data related to co-seismic effect, post-seismic effect, pre-seismic effect and January effect (Start day 1 January 2011).

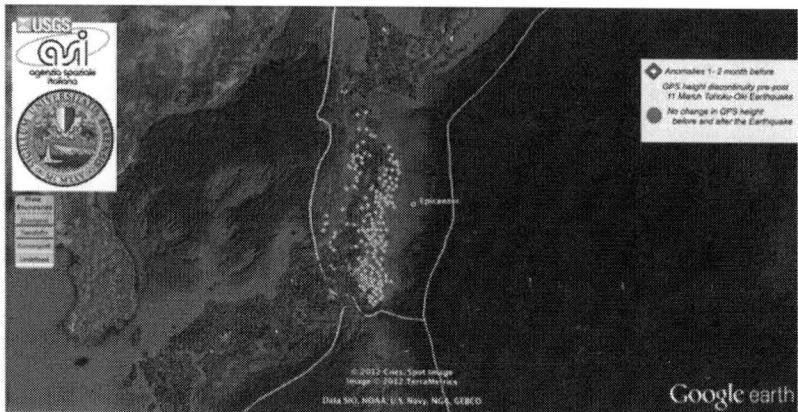

Figure 3. GPS-GSI daily data classified in terms of co-seismic effect (purple), post-seismic effect (yellow) and January anomalies (red-square).

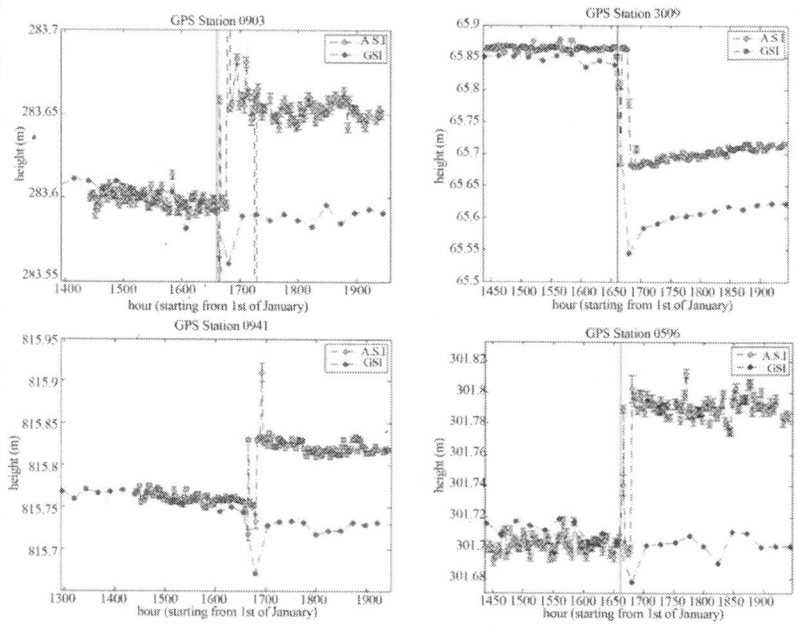

Figure 4. ASI and GSI Complete time series comparison.

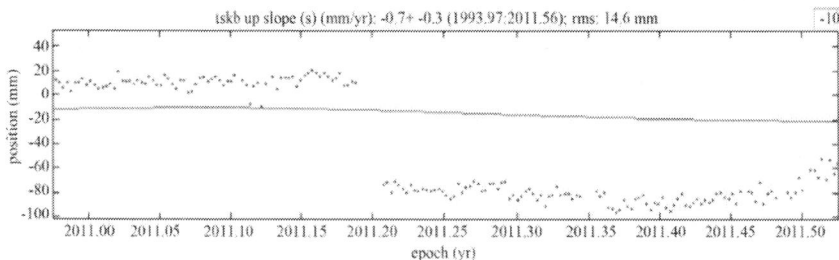

Figure 5. TSKB Altitude time series (JPL-SOPAC data [12]).

GSI and ASI GPS elaborated data substantially agree for what concerns data trends accordingly to the fact that GSI product is a data mediated over 24 hours while ASI product is a 12 hours window with a sliding window of 3 hours.

Only pre-seismic time series have been plotted in order to underline variation (Figure 6), trend fit has been added using a non linear least square method with a function $x \cdot a + c$ a and c are computed parameters. Purple line indicates time when the earthquake occurred.

A variation cannot be seen clearly for all stations, a different plot has been used for a better visualization (Figure 7).

It can be noticed that the 2 cm data dispersion and data trend of ASI elaborated data, substantially agrees with

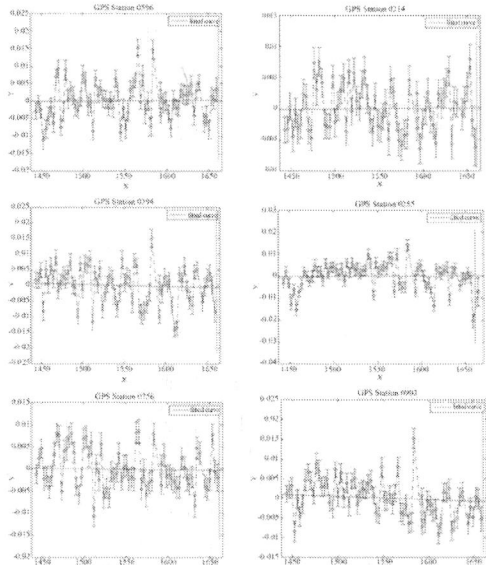

Figure 6. Pre-seismic time series of all GPS stations.

daily GSI data (Figure 8) in fact both data record a decline rate of 29.3 cm/year. According to the target, hourly analysis reduce RMS error from ±28.55 to ±8.63 cm/ year giving statistical evidence to the pre-seismic decline rate.

The explanation of the not so evident pre-seismic decrease can be found in three equally important points:

- Only 2 stations (940 and 903 Figure 7) of the GPS network chosen for 3 hours sliding window analysis show possible pre-seismic anomalies in daily data,

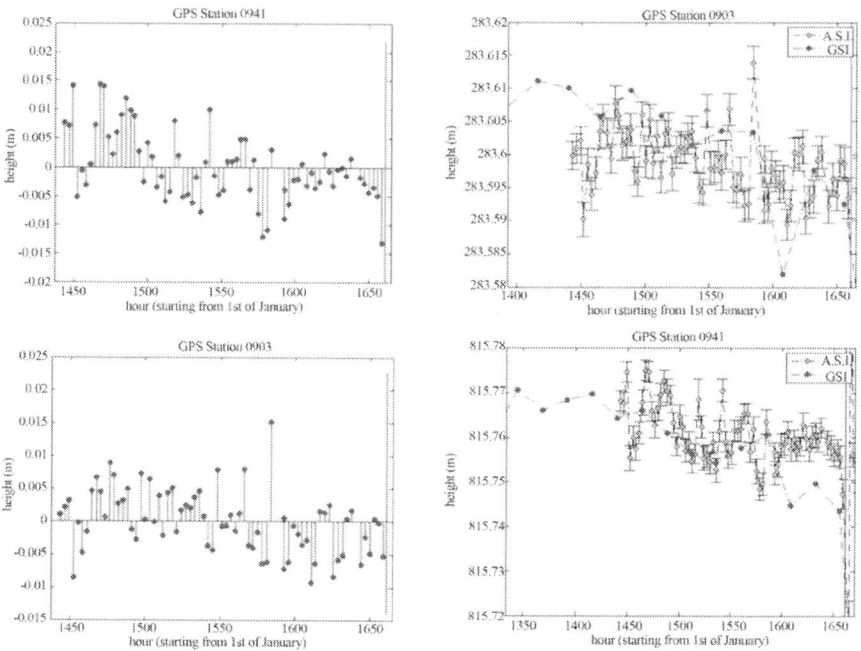

Figure 7. 903, 941 GPS stations pre seismic time series, GIS, A.S.I. Data comparison.

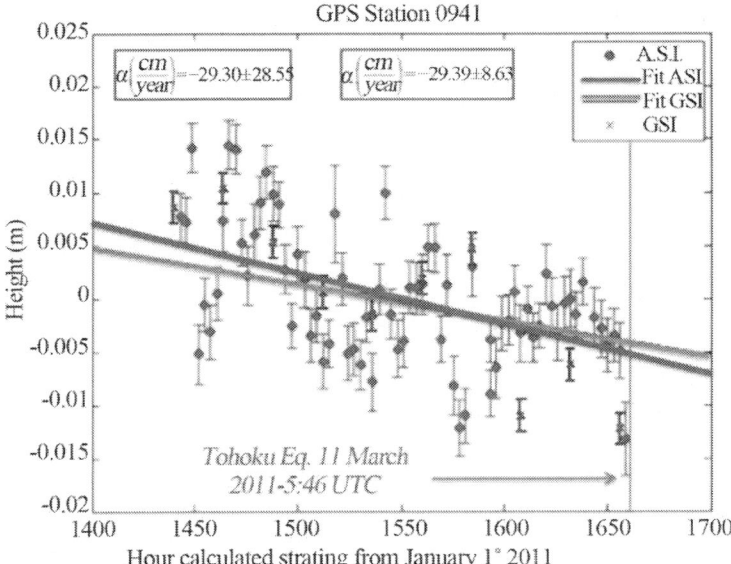

Figure 8. Dispersion and data trend of ASI elaborated data.

- this negative trend is confirmed in hourly analysis with the same order of magnitude.
- Hourly analysis seems to be affected by a higher noise due to the 12 hour window used, GSI daily data use a 24 hours window, this means that if GPS acquires a signal every 30 seconds, each value of altitude is averaged over 1440 more data than a single hourly altitude value. This inevitably introduces a certain amount of noise.
- Hourly analysis covers a time period of only 10 days before the earthquake, this could have hidden in part the decrease effect as in the case of stations 903 and 941 (Figure 8).

CONCLUSIONS

This study has confirmed that an earthquake occurrence is a complex phenomenon that, apart from obvious coseismic and postseimic effects, implies also preseismic effects that can be described as anomalies. In this paper, the anomaly investigated refers to the value of the crustal heights measured by means of GPS techniques. In accordance with other types of anomalies, such as radio [3,5, 13,14] or chemical anomalies, the presesimic effects are mainly located around the epicentral area.

We can conclude that GPS techniques can be of extreme importance in the task of earthquake prediction, and together with other type of anomalies related to seismic activity, we can have the basis for a multi-parametric approach in earthquake prediction.

REFERENCES

1. O. A. Molchanov, et al., "Precursory Effects in the Subionospheric VLF Signals for the Kobe Earthquake, Physics of the Earth and Planetary Interiors," 1998.
2. Hayakawa, et al., "On the Correlation between Ionospheric Perturbations as Detected by Subionospheric VLF/LF Signals and Earthquakes as Characterized by Seismic Intensity," Journal of Atmospheric and Solar-Terrestrial Physics, Vol. 72, No. 13, 2010, pp. 982-987.http://dx.doi.org/10.1016/j.jastp.2010.05.009
3. http://www.izmiran.ru/projects/IK19/
4. Kasahara, et al., "The Ionospheric Perturbations Associated with Asian Earthquakes as Seen from the Subionospheric Propagation from NWC to Japanese Stations," Natural Hazards and Earth System Sciences, Vol. 10, No. 8, 2010, pp. 581-588.http://dx.doi.org/10.5194/nhess-10-581-2010
5. Hayakawa, et al., "A Statistical Study on the Correlation between Lower Ionospheric Perturbations as Seen by Subionospheric VLF/LF Propagation and Earthquakes," Journal of Geophysical Research, Vol. 115, No. A9, Article ID: A09305.
6. K. Heki, "Ionospheric Electron Enhancement Preceding the 2011 Tohoku-Oki Earthquake," Geophysical Research Letters, Vol. 38, No. 8, Article ID: L17312.
7. http://www.gsi.go.jp/
8. http://acc.igs.org
9. http://www.denshi.e.kaiyodai.ac.jp/ja/
10. Bernese, "GPS Software Reference Manual," University of Bern-Version 5.0, 2007.
11. http://facility.unavco.org
12. http://sopac.ucsd.edu/
13. M. Hayakawa, et al., "A Possible Precursor to the 2011 3.11 Japan Earthquake: Ionospheric Perturbations as Seen by Subionospheric VLF/LF Propagation. (in Phase of Pubblication)," 2011.

14. Biagi, et al., "Anomalies Observed in VLF and LF Radio Signals on the Occasion of the Western Turkey Earthquake (M = 5.7) at May 19, 2011 (in phase of pubblication)," 2011.

CITATION

P. Milillo, T. Maggipinto and P. Francesco Biagi, "Variation of Altitude Observed on the Occasion of the Tohoku Earthquake (M = 9.0) Occurred on March 11, 2011," Open Journal of Earthquake Research, Vol. 3 No. 1, 2014, pp. 22-29. doi: 10.4236/ojer.2014.31004.

CHAPTER 6

Acute Psychiatric Trauma Intervention — The January 2010 Haiti Earthquake

Kent Ravenscroft[1]

[1] Psychiatry, Georgetown and George Washington Medical Schools, Washington, DC, USA

INTRODUCTION

Much is written about disaster psychiatry, all essential reading for anyone volunteering to do this kind work. Much less is written about what it is actually like to do trauma work on the ground, often under dire circumstances. How do you actually apply this important body of psychiatric knowledge to traumatized patients and families? And do it when you yourself are anxious, tired, and pushed to the limit? How do you manage your own health and mental health while working in the trenches alongside equally stressed colleagues struggling with flooded clinics, minimal equipment and short supplies? How do you keep your head screwed on straight when the work itself is traumatizing, and the circumstances crazy making? How do you cope as you question your clinical skills and competence, your judgment calls, and your ethical standards at every turn with the circumstances constantly fluctuating? All this happens despite the best efforts of whatever NGO (non-governmental organization) you're working with.

This chapter is about the Haiti earthquake, a disaster of unimaginable proportions. It is about the experience of one psychiatrist, a volunteer with the International Medical Corps (IMC), who arrived soon after the January 12, 2010, 7.1 Richter earthquake. While having volunteer doctors do some direct service, the IMC's major volunteer mission was to train Haitian physicians and nurses to do better ongoing clinical work themselves, both medical and psychiatric. The IMC follows the old adage: Give a person a fish and he eats for a day; teach a person to fish and he eats for a lifetime.

Of course, every volunteer physician or nurse, every psychiatrist, brings his or her own 'baggage', own discipline and experience, own strengths and vulnerabilities, resulting in a range of unique volunteer experiences. Yet there is a common thread to all of this, weaving a story worth telling for future volunteers to consider. By walking in another volunteer's moccasins doing Haiti psychiatric disaster work, you can better prepare yourself for what lies ahead as you embark on your own volunteer effort. Hopefully this will lessen your culture shock, improve your clinical skills, and deepen your satisfaction as you do this demanding work.

While surgeons and medical doctors have certain defenses allowing them to do their arduous trauma work, these same (necessary) self-protections make them variably more immune or less aware of certain other things. On the other hand, psychiatrists, to do their intimate emotional work with trauma victims, have to let down their own guard and become more open to their patients' and their own inner experience, potentially putting them at greater risk in disaster situations. But this openness also gives psychiatrists unique perspectives. These insights can be useful in disaster work in general. Embedded in the ongoing narrative of this chapter are most of the principles and practices of disaster psychiatry. The specific approaches presented represent one particular practitioner's way of doing things, guided by IMC principles; but general principles and application techniques emerge around a wide range of gripping cases.

At heart, this is the remarkable story of a seasoned older psychiatrist who once lived in Haiti as a young Yale undergraduate anthropologist, returning now to spend a grueling month on the front lines of the Haitian relief effort—an experience confronting him with unexpected medical and personal challenges, harbored in long-forgotten buried corners of his own mind.

By offering a candid first-hand description of his time in Haiti—making us feel we are there in the tent and the clinic with him—he provides a memorable basis for understanding and applying the principles of disaster psychiatry.

His journey begins in France.

TUESDAY, FEBRUARY 16, 2010:
FOIE GRAS AND FATE

I was sitting there quite alone at a crowded table when the call came. We were in the Dordogne enjoying foie gras and truffles. I was trying to forget the earthquake that had ravaged my beloved Haiti. I didn't want to ruin the *Les Liaisons Delicieuses* trip my wife had worked on so hard to create. But some 50 years ago, as a Yale undergraduate anthropologist I had lived in Haiti with a voodoo priest and his family just outside Leogane, now the epicenter of the quake. I had kept in touch with my friends there, recently receiving first hand reports of massive death and destruction. Sad and guilty, I feared I would never be asked to join the medical relief effort, wondering if it was because I was now 70.

My French cell phone vibrated in my pocket, jolting me back to reality. Embarrassed, I turned and cupped my hand over the phone. "Who is it?"

> *"Drs. Lynne Jones and Peter Hughes from Haiti with the International Medical Corp. Do you have Skype?"*

> *"Yes, down in my room."*

> *"Could we call you back in a few minutes? What's your Skype name?"*

Twenty minutes later I reappeared at the table, after an interview about my psychiatric background and perspectives on disaster psychiatry. They had actually quizzed me on how I would handle the case of a dazed incoherent Haitian woman found naked walking the streets of Port-au-Prince.

> *"Why are you looking so ashen, Kent?" my wife asked me. 'What just happened?"*

> *"They want me to come to Haiti. They're starting up mental health teams near Leogane in their emergency mobile medical clinics."*

> *"I thought that's just what you wanted?" Rod Drake, my closest Washington psychiatric friend, threw in.*

"I never thought it would actually happen." Then I recalled my surgeon brother-in-law, Mike Ribaudo's warning, "Be careful what you wish for." Well here it was, the die was caste. I was about to face the biggest challenge of my life. Apparently, the Washington Psychiatric Society and World Psychiatric Association had sent my resume on to the International Medical Corp.

There was a hitch, though. After being processed to go, I received an email from the IMC deployment officer saying I had been put on extended hold because of overstaffing. Hurt and miffed, I emailed back an impassioned rejoinder, "Take me in the next few days or I'll go elsewhere!" I didn't want to miss my chance. I also copied Lynne Jones.

Lynne fired back immediately, "Hold on Kent, this isn't coming from me. Let me see what I can do." Within one hour I was off hold and back on track--my first taste of what a brilliant bulldog Lynne could be. She's a British Child Psychiatrist, and a great clinician, teacher, and administrator. I soon learned nobody messes with her. *Thank god somebody like her is down there in Haiti,* I thought.

THURSDAY, MARCH 4, 2010: PARIS TO PORT-AU-PRINCE: TIME WARP

During my check-in for departure from Paris, Air France charged me $300 overweight for my medical supplies and equipment. I even reminded them Haiti was their former colony to no avail. Aboard the plane, seated on my left was a French sapeur pompier, with his large fire and rescue team behind him, ready to do rescue and water purification. On my right was one of the top people in WHO who told me that within the first two weeks there he had to coordinate 240 humanitarian groups, medical and otherwise. He was returning after a break to deal with a much larger number, though some were pulling out with the most acute phase ending.

Just before I left, my sister-in-law Polly, a nurse, talked to her colleague on the hospital ship Comfort, moored off the shore of Haiti, who told her, "We've been swamped with the worst cases I've ever seen, many dying, but many saved. I had one little girl with a horribly infected face, worms and maggots crawling out of her festering wounds. Polly, we've never seen anything like this. No war zone compares. But, you know, I've been moved to tears by the strength and spirit of the Haitians I've seen." Just as I was packing up, Patti told me about the conversation with Polly, then looked me straight in the eye, "Kent, do you really know what you're getting yourself into?" My anxiety shot sky high, "Please, that's enough. I have too much to think about already."

For me, at least consciously, I was more worried about my so-so French, my rusty Creole, and my ability to help psychiatrically in the midst of such devastating physical tragedy. Finally, I said, "Going down to Haiti feels like the biggest final exam I've ever taken. I thought I was completely finished with things like this." I have been plagued by performance anxiety all my life, and was glad to be done with it. "I'm not worried about how you'll do," Patti said. "I'm worried about your health and survival." "What about your pulmonary embolus three years ago. And the Coumadin you're still on?" "It'll be okay." I said, "My health's pretty good now. Stop being so anxious." "Me? Look at your hands, Kent." We both stared down at them. They were trembling. "I've always had that hereditary tremor, nothing new. You know that's why I didn't go into surgery." "Come on. You're vibrating like a tuning fork." "Okay, so I'm a little anxious, but I'm going. I have to. You know why. And I need your support." "Just be very very careful," she murmured.

As the plane for Haiti took off, my cabin conversation slowly gave way to personal reflections. I slipped on earphones, and leaned back. A subtle sadness filled me. Intimations of the loss of the Haiti I knew 50 years ago came into my mind. I had experienced Haiti back then in the youthful blush of naive enthusiasm and brash denial, allowing me to tour carefree all over the country. Now TV images of collapsed buildings and survivors being hauled from rubble floated through my mind. I had glimpses of what I would be dealing with once I arrived. And I would be seeing it from my sobered vantage point now later in life. Even so, I felt a shudder ripple through me. Something more was troubling me. Something deeper. Just then, "Haiti Cherie" came over my earphones, taking me back to glittering evenings dancing at the posh Petionville country club. I was able to push more depressing thoughts away with these pleasant memories.

My first summer in Haiti had been such a high, the stuff of dreams. I was in love and loved what I was doing. I chronicled everything, capturing it all in love letters to my girlfriend Linda. I had met Maya Deren in Greenwich Village as a Yale junior. She had lived with a Voodoo priest, Isnard, during her Guggenheim to study Haitian Vodun dance, and authored the fascinating book The Divine Horseman: The Living Gods of Haiti. My hope was to study Voodoo spirit possession and try to understand its behavioral content and psychodynamics. The project took on such major proportions I applied for Yale's Scholar of the House Program, freeing me from all class work save presenting my research work monthly, and writing my Scholar of the House thesis. When they accepted me I was ecstatic. I was still doing my pre-med courses on the side. Memories kept filling my mind.

Someone tapped my shoulder snapping me out of my reverie. "Would you like a snack?" the stewardess asked? "No, I'm having a delicious time as it is," I said, slipping immediately back into my daydreams. Then I fell asleep. In my dream from the corner of my eye I thought I saw something black and hairy crawling onto my left shoulder. I screamed and sat bolt upright, wrenching myself away from it. I scared the hell out of the French fireman next to me. "Hey, buddy, you all right?" "Sorry," I said. "Must have been a nightmare or something." I had also bumped the WHO guy next to me and he was eying me suspiciously. "I'll be okay. Just a bad memory." I settled back into my seat, not at all sure I wanted to close my eyes again. Haiti was rapidly coming closer and my second summer there had been a nightmare, including two encounters with the tarantulas living in my room. Momentarily I put my psychiatric hat back on, realizing I was having a return of my post-traumatic stress disorder. As our jet approached Hispaniola, I found myself wishing we were in that old slow prop-driven plane of yore. I needed time. I had been enthusiastic about going, despite my anxiety. But the reality was fast approaching. Was I equal to it?

When the devastating 7.1 magnitude earthquake hit Haiti in January, because I had been an undergraduate anthropologist in Haiti 50 years earlier, my deep love for her people moved me to volunteer. Back then, I was an undergraduate living with a voodoo priest and his family outside of Leogane (now the epicenter of the quake), doing research on voodoo spirit possession, contributing to the theory of multiple personality. Now I was a 70-year-old retired psychoanalyst and child psychiatrist, facing a month's tour of duty with the International Medical Corps. Soon after I arrived they sent me to Petit Goave just beyond Leogane, to form a mental health team serving the five IMC mobile medical clinics (including one boat clinic for a fishermen's village). 70 to 80% of area houses were damaged or destroyed, killing large numbers of men, women and children. Surrounded by death, destruction and dislocation, the surviving people constructed makeshift tent cities everywhere, with donated tents finally arriving later. Survivors from the Capitol poured steadily into this outlying area, aid lagging far behind.

TRAINING MISSION

Soon after I arrived, I found out what the International Medical Corp had in mind for me. Our IMC mission, while providing urgent psychiatric care, was aimed primarily at training clinic Haitian family doctors and nurses to do independent sustainable psychiatric assessment and treatment. IMC emphasizes water, food, shelter, community, and security as essential to recovering mental health. Our psychosocial clinics stressed emergency psychiatric intervention techniques designed to help people deal with mass trauma, dislocation, loss, grief, anxiety and depression, as well as seizures and acute and chronic psychotic illness. We also taught a basic psychiatric pharmacy. Like military psychiatry, we focused on interrupting stress, anxiety, phobic and depression-based symptoms interfering with normal grieving and self-righting after mass trauma—linked with strengthening of family, friends and community

My personal mission was to create a mental health team that traveled weekly to each of five medical clinics, providing clinical teaching for the clinic doctors while we gave direct patient care. I also had to give two Saturday trauma psychiatry lectures and workshops to the Haitian clinic doctors and nurses—a daunting task since they were on the front line and I hadn't had any experience with them or their patients yet. Right after arriving, I gave my first workshop and lecture. After a sleepless night of frantic preparation, I took a deep breath and waded in.

TRAUMA PSYCHIATRY SEMINARS

In our Saturday Seminar we taught 12 Haitian doctors and 19 nurses from our 5 clinics, plus residents from Notre Dame hospital. I did a group exercise teaching them relaxation and imagery techniques to interrupt cycles of anxiety and repetitive thoughts, and then while in the relaxed state, had them visualize where they were when the earthquake struck, helping them recapture and work on their own inner experience, to increase empathy for their patients. I stressed cost-effective front line stress reduction nursing group sessions for people they identify--just a session or two, using techniques I was modeling with them.

In my lecture, I stressed the importance of their health care presence at the clinics which were strategically placed in or near the tent camps and the destroyed villages, reinforcing the impact of their caring presence and caring activities, their laying on of hands, their quick but careful exams, including their mental health first aid and triage. As front line workers, their work and reassurance gave hope and momentum to recovery for this vast impacted impoverished, yet strong resilient group of Haitian people, helping them move along in their expectable stages of recovery from a mass disaster. I emphasized that they should think of how to normalize their patients' 'abnormal' experiences, maybe based on their own seminar guided-imagery experience I had provided for them earlier in the day. This might help correct their own and their patients' dire and fearful self-diagnoses, stirred up by their new 'symptoms', their expectable range of new weird thoughts and feelings. They needed to be careful not to overly 'pathologize' what their patients presented by becoming familiar with their own experience, by coming to know the normal stages of mass disaster recovery. This would help themselves and their patients avoid getting stuck and becoming chronically symptomatic.

I emphasized the individual and group importance of their presence at their Clinics, that they were, just by their presence, and their laying on of hands and manner of caring, a transference object of great importance for the tent camp and village they were in, becoming a healing beacon in a troubled mental sea. Although I knew they felt guilty about the long lines every day, and their brief problem-focused encounter with each patient, I said that their Haitian patients were used to waiting for care, care like they had never had before, and that even waiting in the clinic near their doctors was curative as part of the placebo effect. Just knowing the Clinic was there mentally, and spreading the word to the camp or village, was good for people. I repeatedly emphasized that they were on stage for their patients, as they delivered care in these open tents with every one watching, and that they should never underestimate the importance of using themselves as a powerful part of the healing. Patients who sit and wait are in a healing presence that sets the mental stage and also cures in its own right.

IMC also stressed how much shelter, food, water, security and reunification or 'retribalization' means (including connecting people with religious and secular groups if their own families are shattered or dead) in providing the substance and holding context for their recovery from mass disaster. Several nurses and doctors shared their own family experiences, and then, feeling safer and more encouraged, their patient experiences. Then I looked directly at them and asked them how they themselves were doing, asking for a show of hands about how many had lost family members, how many had their houses destroyed, how many were living in tents outside their houses, and how many in tents in the camps.

Though many were more fortunate than others, all were traumatized, and about a third had been deeply affected one way or another, many with significant losses, several now living in the camps. When the subject of tents came up, I noticed two nurses looking down and huddling privately. I finally asked if they could share what was going on. With some embarrassment but plucky honesty, one nurse confessed she didn't even have a tent yet and was living outside with family members in one of the camps, grateful the rains hadn't come, and proud she made it to the clinic every day to work, somehow looking clean and kempt. It seems tents are now in short supply in Haiti and there is still great need.

I had actually volunteered to live in a tent when inside beds in our IMC residence were in short supply, but now I suddenly felt guilty. One nurse knew about it and spread the word in the audience while I was talking about all this. I heard muffled laughter and finally asked what was going on. The nurse finally confessed, adding, 'So you're in the tent city just like us!" I smiled and shook my head. Everyone laughed.

I also stressed that my team role and mission, in addition to helping provide direct urgently needed psychiatric care, was aimed at training Haitian family doctors and nurses to upgrade their mental health skills. We all agreed they wanted to become more adept and independent in this area, wanting to improve their own psychiatric assessment and treatment work in their medical clinics.

During the lunch break, as everyone else made a mad dash to be first in the buffet line, one young doctor came up to me and said he was unsure about the purpose and usefulness of the imagery recall exercise. When I explained it again, his eyes rimmed with tears. He told me about pulling children, some dead, some gravely injured, from under crumbled concrete slabs in the house next to his after he and his kids managed to get out safely--just before his own house finally collapsed completely. We talked at length, he was grateful, and I thanked him for having the courage to talk with me.

I knew I'd be working with him soon in one of my weekly clinic rotations, and was deeply moved by his experience, and by the group at the conference. All this certainly broke the ice for me, and I hope for them. I felt poised, ready to go out to meet them on the front line, working along side them, treating their patients with them. At that point, though, I was all enthusiastic readiness and no real experience. Unless you count working in a Washington, DC, inner city emergency room.

How would I do? I had a strange thought. I should summon up voodoo god *Maitress Erzulie Gran Freda, and say, "Please come up from the abyss. Be my divine guide. Help me to find the healing words I need"*. Where did this come from? Then I recalled my last day in Haiti 50 years ago. The voodoo priest I was living with had taken me aside, into the inner sanctum of his temple. He went into a trance, and his Maitress Erzulie possessed him. In her characteristic deep gravely voice she said, "You have already sewed fingers back together, and treated our TB and dysentery, so for us you are already a doctor. But some day after you finish your medical school, you will once again be called upon to come back to Haiti to serve us again in our hour of greatest need. Her words comforted me even as a chill ran down my spine.

Now that day had come.

CLINICAL TRAINING AND TREATMENT

Petit Guinée Clinic

Let me tell you about my initial days in one of IMC's clinics, the Petit Guinée Clinic, to give you a taste of my experience.

As we bounced along in the van, one doctor quoted a most recent CNN commentary, which said the rubble from the Haiti earthquake would fill the entire Washington Mall to the height of the Washington Monument. My heart caught in my throat as a realized with fresh impact what they were seeing. As we drove along, we saw a house totally destroyed, with a slanting slab of roof, now taken over by goats standing at the peak. At least they wouldn't be eaten at night, unlike the 'free-range' chickens with nowhere to hide. I now had more sympathy for the roosters and realized why they were crowing at all hours. Packs of hungry dogs roamed the night, seeking whatever they could scavenge, given the scarcity of food and leftovers.

On the way to Petit Guinée we drove through the poorest section of Petit Goave, a beautiful seaside location, but also one of the hardest hit. As we arrive, the Petit Guinée patients are registering as Drs. Affricot and Louis confer with nurses.

I was privileged to see how the staff set up the clinic. Patients were already sitting on fractured cinder blocks for stools, squatting in classic 'Haitian style' all around the periphery, perhaps 75 people strong. At Petit Guinée Clinic, some mothers were breast-feeding, other mothers and fathers holding sleeping children, all eager but respectfully waiting for a turn--all huddled under a huge, slightly twisted corrugated roof with open sides. Tables were set up, and blankets suspended and tied into makeshift walls, giving a semblance of rooms and privacy. Chairs were at a premium, as were tables. I had worried about how things would be set up, so my nightmare wasn't in vain. We came well prepared, all very useful as the clinic began to roll. I was given a corner up on a cement dais, a remnant of some sort of stage with a shiny pole in the center and an old bandstand. At that point I was oblivious to the former nightclub we were working in.

Dr. Affricot was chief of the clinic that day. So there I was on stage for my first teaching clinic. Without a drum roll we saw our first patient.

Pierre, a shy, taciturn eleven-year-old, was sleepless, constantly hearing the cries of a baby, and the voices of dead neighbors. He had been holding a neighbor's baby when his house collapsed on him. His mother could only see his head when she tried to rescue him. He tried to protect the baby in his arms, but it was gasping when they got him out. The baby died on the way to the hospital, crushed in his arms. He felt horrendously guilty, not helped by the baby's angry grieving parents, whose house had also collapsed. His mother added they weren't really talking about him personally, but he felt guilty, even for surviving. He had had a friend die 3 years earlier and heard his voice for a long time, thinking at times he even saw him in groups of children, until taking a second careful look.

I worked with the doctor to do the interview, using the interpreter to get the story details and give feedback, providing guidance, at times even speaking in my rusty Creole to the boy and his mother. I said he had made it through the mourning of his previous friend. He had more experience with this kind of thing than most kids. We told him he had more complicated grief work to do this time, but by past experience had what it took to work his way through this one too. We said we felt he would do fine. We told him and his mother he was doing too much emotional work in his sleep at night, giving him bad dreams and sleepless dreamy voices during the day. I explained they needed to bring this into the daylight when he could do more effective emotional work.

The mother was advised to have a little session with him each evening to gather all his worries into her mind and arms, helping him clean and clear his mind, reassuring him she would work on them for him so he could sleep—kind of like what you do with Guatemalan worry dolls. She should also tell him he had done all he could for the baby. Nobody could have protected him more, not even his parents. Because he was a shy boy with a strong conscience, making him very self-critical, she needed to tell him to ease up on himself. We gave no meds, but rather a follow-up, saying we felt they would be a good team during their healthy grief homework. They left encouraged and armed with active self-help they could carry with them.

The next woman had severe palpitations. She was on the way home when the earthquake hit, seeing friends in front of their destroyed houses wailing for dead or missing children. She rushed to see how her 5 children had done, finding 4 alive in front of their collapsed home, '*Grace a Dieu*'. But her 5th had not made it home from the school, which had partially collapsed. She wanted to rush out to find her, but her children reassured her she would come home. And she finally did, with stories of other kids being hurt or trapped. It was 3 days later, a delayed reaction, that her short agonizing vigil waiting for her daughter triggered severe palpitations.

She had pre-existing high blood pressure, was on a medication, and worried her heart was giving out, with bursts of rapid beating (palpitations) making her feel she was dying. She let us know she was helping many of her grieving friends, and felt her heart problem was physical. But she had never had this before, except slightly walking up steep hills.

It became clear, after taking her blood pressure and taking her pulse rate, and listening to her heart, that she was physically okay, though we agreed she should see her doctor to get checked out, maybe even have an electrocardiogram. But I told her we knew what was going on, and that she had the strength and intelligence to work this out, letting time and simple techniques restore her trust in her body and in life. I explained the endocrine fear response and her tendency to make scary self-diagnoses escalating her panic. We noted her previous fast walking would make anyone's heart beat faster, and that the new bursts of heartbeats were different, a normal fear response that had gotten a little stuck. To deal with that she needed a couple of techniques to counter thoughts or noises, or after shocks when they triggered them. We taught her to blow into a sack and the Valsalva maneuver, like when you bear down to grunt at toilet, or during childbirth.

The Valsalva causes a neurological reflex (vaso-vagal) that slows the heart. You may have read about this in diving mammals. We told her to use it to interrupt the beginning palpitations. This both works and is a cognitive distraction. Just knowing you could take control helps a lot. We also showed her the paper bag sealed around the mouth re-breathing technique used for hyperventilators, but also for palpitations. The 'sack' re-breathing technique decreases O2 and increases CO2, and distracts— decreasing her overbreathing and tingling and dizziness caused by excessive breathing and increased O2 levels. We also urged her to be a smart scientist, noting down obsessively each time she had such an attack, so she could outfox the triggers, and disconnect them with an 'I told you so", just as she could help her friend do. She needed to be a kind doctor, and not scare herself. She got the hang of it, and understood the psychology and physiology of it. She was a schoolteacher so I suggested she could help teach this to scared symptomatic friends, as she herself got good at it. By becoming active and masterful and a 'trauma recovery teacher', she could help everyone.

Another patient was glassy-eyed and depressed, showing us a certificate of scholastic accomplishment earned by her 21 year old son. His handsome picture smiled up at us from the certificate. Between sobs she told how he was teaching in Gressier, away from home for a while, and was crushed in his little room by the earthquake near his school. She was almost inconsolable, in deep prolonged, but not 'arrested' mourning, yet was bordering on depression. We listened with near reverent attention, checked on her friendship and religious network, and noted she had high blood pressure. She also had serious insomnia. I suggested they add Atenolol, both a relaxing sleep promoting and anti-hypertensive medication, hoping to help her through this sad, sad passing. She had other children to live for, but we would follow her up closely next week just to make sure. We felt we needed to keep her alive for all of our sakes, and asked an accompanying friend to keep checking in or her and to bring her back for her next appointment.

We saw other patients, and as time went on I relied on the Haitian doctor more, since we were hoping to give our clinic doctors increased front line competence, a good sense of basic psychotropic meds, and diagnostic acumen for triage—-and referral, if absolutely necessary. But where could we refer? The psychiatric hospitals were mostly destroyed or seriously overcrowded and understaffed. There wasn't much psychiatric care to go around, and most people, even if deeply affected, were, with simple help, resilient and self-righting—if they had their basic needs met, that is, shelter, water, food, and security, plus some social connectedness. We had to do what we could, enlist friends and community, and just tolerate the uncertainty as best we could.

We had one other woman in the Clinic today who lost one child, an aunt, and her house. Her business establishment also collapsed and then was looted, and, her van was trapped under a concrete wall. So she and her family were without even a tent and no means of livelihood, after enjoying a comfortable, productive middle class lifestyle. She was depressed, and, I sensed, quite angry underneath. Because of this and a sense of pride, she was unable to reconnect with, and in fact avoiding, her Pentecostal Church. She seemed close to needing antidepressant medication, but we gave her a light sleep med at this stage. She and her IMC Haitian doctor preferred it this way. He pointed out to me one had to be on costly antidepressants a long time, and we sensed she might come around the corner if we waited. We planned to see her again next week, just to make sure. Taking the mental pulse and providing close follow-up were the key. We didn't want her remaining children to suffer a maternal suicide because we were too conservative and cost-conscious, given everything else. But for her this proved a wise choice. Her doctor had really connected with her, they were on the same wave-length, and later I would see her coming back to see him.

Mirogane Clinic

Now I'll tell you about another day, at the Mirogane Clinic. Mirogoane is an hour by van from my base in Petit Goave, over a road full of ruts, the roadside lined at times by beautiful coconut and banana trees and sugar cane fields, with a gorgeous backdrop of crinkled denuded mountains, a deceptive sprinkling of green scrub growth in a few places. The trees are scarce, mostly cut down for cooking charcoal due to overpopulation pressure. Looking ahead from the van I see throngs of people, collapsed buildings, goats attacking burning refuse for fruit peels. Gaily panted trucks and cycles hurtle toward us, with tent cities rushing by on either side. The road is periodically scared by those zigzag crevasses, and deep cleavage drop offs, stunning reminders of earthquake forces scarring our poor Haiti, each mercifully requiring us to stop our headlong rush from time to time—the new Haitian equivalent of speed bumps. Speaking of these, because tent cities are all along these mostly mud roads, and the dust horrendous, people in these roadside cities create big, makeshift mud speed bumps. Only the animated conversation with some volunteer

Hopkins Docs, assigned for the day and Nurse made me forget the life, and death, teaming around us.

Pulling up to Mirogoane Clinic, I haul my red backpack up the steps to a clinic bursting at the seams with Haitians, all camping out patiently in anticipation. I had been told there would be plenty of chairs for my mental health clinic in Mirogane, only to find they were scarce. They tried to put me out among the people in plain air, but I scouted around and found a cramped back room, moving soiled instruments and half-empty bottles of medicine and antiseptic out of the way. Finally, I scrounged up three bent rusted chairs and a bench. I was in heaven, and in business. Except no ceiling. The wall went up only 9 feet to a high airspace transmitting the hubbub from the next room. Good for ventilation but not privacy. And there was no door.

After getting oriented with my Haitian doctor, Dr. George, our first patient walked in, referred by one of the Hopkins gals. Suffering from earthquake losses, and quake shock anxiety deepened by aftershocks, we prescribed her some Diazepam and anxiety reducing exercises which we demonstrated (progressive relaxation and deep breathing techniques). We added the 3-breath-technique, where you take a deep breath, hold, then take it deeper, hold, and then as full as you can and hold. You do this three times, concentrating on your breathing. Doing this is incompatible with remaining anxious. We also prescribed homework telling her to share her story with family and friends to reconnect her with her own experience narrative.

In the midst of this, a toothless wizened old man, drunk as a coot, came rolling into the room giving us all high 5's. He was to be our next patient, but inebriated and high, he had jumped the gun--his poor impulse control written large in the breeze, along with his rancid alcoholic breath. We saw him next. He was infectiously delightful, all a sad deception. As he raved on, a tear dropped from one eye during a fleeting mention of losing a family member, covered immediately by gay word torrents. He told us he had been drunk most of the time for 8 years, and that it was his sister's fault. She had been a raging alcoholic before him, until she saw a Voodoo priest who, for a sizeable fee, removed the "devil drink" from her, unfortunately putting him in our patient. I helped Dr. George accept this story with a straight face and explore the personal causes and all the awful sequelae of such chronic drinking (black outs, the DT's [Delirium Tremens, famed for its kaleidoscopic 'pink elephant' hallucinations] and Wernicke-Korsakoff syndrome [with its loss of memory and sincere confabulation], etc). Miraculously, he had been spared by the earthquake and lived to tell his story. I had the doctor do a careful mental status to see if our patient, besides intoxication, had the hint of other brain damage. He seemed pretty clean, to our surprise. We were naively hopeful. But hope is important for everyone in this kind of situation, including—or especially—for us treating doctors.

Then I asked if Dr. George had seen the patient's brief moment of tearing up. He recalled it, so I asked him to explore what lay behind this fleeting hint. A lot of underlying isolation and sadness emerged, which the patient usually camouflaged by his "hail fellow well met" veneer. Picking up on his sister's exorcism, I said to him I knew about Voodoo, and had thought of taking training myself for the *Ason*(the priesthood). With a knowing smile, I said we would be willing to receive his 'drink devil' if he wished to give it up to us. But, we said, we could not give him a proper examination for diagnosis and a path toward cure unless he were sober. Looking him in the eye, Dr. George asked him if he could try being sober for the upcoming week, to get his body and mind ready for our next visit? The patient's eyes grew wide, then narrowed, and he accepted the challenge. This was based on our sympathizing with his underlying loneliness, which had touched him deeply. Without challenging his beliefs, we told him we felt he was taking the wrong medication, his self-prescribed alcohol. He agreed to come back the next week. Chronic alcoholics are tough, especially in this environment, and yet this guy had pluck, and the obvious available mental hook revealed by his tears. But we would have to wait and see.

Then this lovely healthy-looking but somber young woman, 21 year old walked in, complaining of insomnia, palpitations, visions and voices, but of a very particular kind. The voices and faces were fellow medical students and young nurses who had been trapped together with her in the basement as their building collapsed on top of them, there in Port-au-Prince. Trapped in pitch-blackness, pinned under rubble, she could hear the voices, the screams and cries, of those injured and dying all around her. Over four grueling days she heard these voices, voices of her friends, holding their faces in mind to keep herself going, only to hear those voices becoming fainter and weaker, and finally dying out, leaving her alone with only one friend's voice, somewhere way up above her. This faithful friend knew she was down there somewhere below her, and told the rescuers, when she heard them above her on the third day, saying she was alive down there somewhere below her, guiding them in both their directions. Then her voice, too, became weaker, and died out somewhere above her, leaving our patient utterly alone.

As rescuers got closer to our patient, at first her own voice was too weak to call out on her own behalf, though she could hear them calling her name. Finally she found the strength and called just once, loud enough to be heard. But the rescuers found that the pieces of concrete over her were too big to be moved. They told her they wouldn't give up, but they feared they wouldn't be able to do anything soon enough, telling her she should hold on as long as she could, and they would do their best. She lost hope though, hearing rescuer voices growing faint above her, as she hung in darkness, her sense of day or night completely lost. Her throat was parched, and her loneliness deafening, but she didn't give up. She was the last of all her pre-med friends to survive. Then, finally someone got to her feet. We found out, at that point in her story, that she had been suspended upside down the whole time. As she talked with us, encouraged by us to open up her darkest hours, her voice grew stronger, calmer, and more certain. I finally blurted out I was so proud to have someone like her becoming a member of our profession.

She broke out in a radiant smile, and told us she was hoping to go back to medical school when classes started. She would be finding out the next day when that would be. As she said this, her voice quavered and got softer. She already knew two thirds of her class of 45 had died, and confessed she was petrified about going back. She was having palpitations and hyperventilation, with near panic attacks whenever she thought about getting near the collapsed medical school building again. She dreaded finding out if even more had died, and wondered about the teachers. Dr. George and I gave her some diazepam to take the edge off her anxiety, and help with her insomnia, and gave her three desensitization and behavioral techniques which would give her ways to systematically move toward mastering her feelings of fear, her foreboding thoughts of impending disaster, and her phobic avoidance of her school and the future.

As we went over these techniques we found out she had been a student leader, and suggested she might be a good teacher and leader for student groups with whom she could share her experience and techniques for their shared anxieties, helping classmates to resolve their symptoms. Facing her own understandable feelings and reactions, using all her robust strength, might allow her to return to her support community and show them how to work together, to resolve their shared fears and losses and accomplish mass mourning. By the end of the session we had a sense she would be able to make it, and help shed light on the darkness they all faced. We asked her to come back with a journal of her homework accomplishments to an appointment at our next clinic, a week hence, if she didn't remain in Port-au-Prince. We told her we all felt she, especially, would be able to make it. We clarified issues around survivor guilt, emphasizing that she was living for herself, and that her self-exploration and healing would allow her to be a fine compassionate doctor sometime quite soon.

As I write this, tears are streaming down my face. This young woman, in particular, takes my breath away and makes coming down here worthwhile. But doing this kind of work is often like doing surgery without benefit of anesthesia, and yet painstakingly important. Many of my young Haitian doctors felt it was mean to have patients remember and feel what had happened—until they saw how it unlocked and freed people to face and re-embrace their lives and their hopes for the future. I didn't have time to decompress while working with our Haitian patients, and didn't discover how much they were affecting me until I try and write about it. I admire the strength of many of the people I see and yet don't want to be too idealistic or naive. Or too optimistic. We do what we can, and hope for the best.

The next patient was a 22-year-old girl, with trouble speaking because of a tight, tremulous aching throat. "Feels like a lump in there," she said. She also had trouble keeping her eyes open when this was bothering her, and dizziness and hyperventilation. All of this had happened, on and off, in the past--around failing to get in medical and then social work school. The symptoms reappeared with the earthquake, coming in waves. It was then that she mentioned her brother had died in the quake, someone she was very close to. The moment she heard of his death, she couldn't open her eyes for hours. Her other brother in Cap Haitian was calling and crying openly, but she couldn't cry at all. My translator Tessier, a teacher, tapped me on the shoulder, "Dr. Kent, I knew her brother, a student and a friend. He was a great guy. I feel so sad to hear my friend died." He had a rim of tears. "What a loss."

I said, "Tessier, tell her what you just told me." When he did, she began to cry, but her throat tightened up, her voice became strangled, and her eyes started to scrunch shut. And the crying stopped suddenly, as if caught in her throat. I said to the Haitian doctor, "See that, she has Globus Hystericus, or anxiety-based laryngospasm. Some people call it the "Stifled Cry Syndrome". She is using hysterical conversion to protect herself from overwhelming loss and sadness, a kind of psychophysiologic spillover into her body from her intolerable broken heart. She needs to close it all out, not feel it and not see it with those scrunched up eyes. When her normal waves of sadness hit her, coming up unexpectedly from her body, she cuts it off by this defense resulting in her symptoms. But she's on the road to recovery if we can help her, or get her family to help her, face her brother's loss and tolerate her mourning."

"Should I tell her all this?" said the Haitian doctor. "By all means, but also include her aunt here, and additional calls with her brother, so they can work together on this. Make it definite homework. We can give her the Sac Rebreathing Technique for her hyperventilation, as well as Relaxation, and Imagery approaches, but her best bet is family help through shared family mourning. Have her come back to see you briefly for several follow-ups. You can really help her a lot with very little."

The next young man, a 27 year old, was a follow-up. For the second time in his life, he had lost close friends, though unscathed himself. But he had ended up not being able to hear very well. Voices seemed far and faint. After carefully surveying his history and hearing situation, we felt he was having hysterical negative auditory hallucinations, basically losing hearing ability because his school had collapsed and, on a deep automatic level, he needed to keep from hearing all the horrible things he had heard. The shouts and screams of those dying or suffering around him, as well as his own petrified unvoiced thoughts. So he virtually gave up hearing all together. He had heard all the voices of fellow students below, the injured screaming in agony—voices he kept hearing in his dreams and waking mind's ear. He was plagued by nightmares, which constantly awakened him. He desperately needed to get rid of all this so he could stand his own mind and not be held hostage, or driven crazy. He wanted to be free to pursue his life. But the cost of his hysterical protection was severe, not being able to hear other things in order not to hear these anguished cries.

In his follow-up visits with Dr. George, this being the third, with support and some anti-anxiety medication, he was already beginning to hear better, and as memories and feelings came back, he was becoming flooded with painful but laudable (valuable rejuvenating) grief. In coaching Dr. George, and trying my hand at a formulation and interpretation in my 'French-Creole', I was able to give him the conceptual tools and his patient the self-empathic support to understand his tortured mental journey. Dr. George gave him hope and courage that the road ahead was possible and the end point 'in-sight'. In was particularly gratifying to work with Dr. George. He was quite sensitive, and also the doctor who had come up to me during my first seminar sharing his own losses. He was open to his patients, their profound grief--tolerating and sharing through his inner empathic resonance.

Dr. George and I discussed the fact that this bright sensitive, fairly timid and inhibited patient had been able successfully to get through similar symptoms with his pre-earthquake loss, which gave us hope, and a predictive model for recovery, and a basis for a hopeful prognosis, though we emphasized he was carrying and bearing a lot more this time. We also discussed the burdens and pitfalls of survivor guilt, and the needed self-maintenance around it. I encouraged his continuing work with Dr. George, discussing his self-protections. I felt his more primitive early defenses were giving way to healthier and healthier ones, now resulting in the painful but constructive waves of grief, and associated guilt, as he dealt with his losses. We urged him not to be ashamed, but to write in his journal and to begin to have the courage to share has experiences with close friends and family. We invited him to come back in a week to help us appreciate all the hard good work he was doing. He left confirmed in his continuing progress. I was proud of his Haitian doctor's work with him, which I shared with him after his patient left.

Our last patient was a cute little girl, 5 years old, with severe developmental delay from birth, who had seizures and had lost her medication when her house caved in, and her doctor had been injured and was unavailable. So she needed to get her two seizure meds from us. We breathed a sigh of relief at such a routine request, which we filled with pleasure. We determined in the process that her medications were not controlling her seizures very well so arranged to adjust them and have her come back until we got them right.

My gifted interpreter, Tessier, a schoolteacher out of work because his school had been damaged (in general the schools were still closed), turned out to know a lot about these patients and their families. After the Clinic he confessed he felt dizzy and drained, and a little sick to his stomach. We both commiserated about all we had heard, agreeing it was a lot to swallow, especially with open hearts and minds. We both needed some R & R. And yet he felt he was privileged, and learning a lot. He also pointed out that the head nurse, who had been at my opening seminar, made it a point to come and sit in on our work. I had a hunch this would pay dividends for her and for the clinic, benefiting future patients. We all looked forward to meeting again next week. I stuck my head in and gave the Hopkins doctor feedback on the patient she had referred to us.

My shirt was drenched by this point. My best self-care, though, was the fact that I had invested in a blow-up camp pillow, which at first I was embarrassed to take out, until my seat couldn't take the rock hard chairs any more. So I would blow it up, soon making it a ritual. Tessier and the doctors, and the watching patients, especially the kids, loved to watch. And boy was it comfortable during those long grueling sessions, where I had to have the quickly spoken Creole translated to me, and then my words fed back in Creole to the doctor, though at the end of each case I tried speaking some in Creole to model how to give interpretations, at times drawing quizzical looks, at other times confirmation of my words. Often, though, Tessier had to re-translate my 'Creole' into Creole. I was happy to be rescued. We kidded each other that he would soon be getting his own psychiatric diploma. I bought a BIG coke from a vendor on the way out which I guzzled thirstily on the long ride home.

My day at the Mirogane Clinic was but one of many over my month in Haiti, each a step along the way of my great challenge, one that brought me great satisfaction, and continuing profound sadness, as we moved along dealing with the Haitian earthquake disaster.

I should mention that Tessier surprised me two weeks later, sharp observer that he was. "Dr. Kent, you remember that alcoholic guy? Well, he came by while you were talking to your last patient, caught my eye, and whispered I should tell you he was doing much better, though he feels sad at times. He didn't want to take up your time. Tessier told him, "Yes, maybe sadder but much wiser and healthier now---keep it up! I'll tell Dr. R for you. And come around to see us each week. Seeing a success like you really encourages the doctors in their work." The patient seemed very pleased with the idea he could help the doctors, and said he would. He asked me to thank Dr. George and you. "And his breath?" I asked. "Not even a whiff of alcohol. He was sober." We both smiled.

SUMMARY OF CASES SEEN

To summarize, I have listed the cases we saw at the Guinée and Mirogane Clinics during just two days of training and treatment. Let me alert you that these cases represent a sample of just two out of twenty clinic sessions I carried out during my four weeks of rotations to five clinics each week. And there were different medical and health personnel at each clinic. While these cases were supposedly selected randomly based on which patient was next in line, I suspect that the nurses made some choices behind the scenes, knowing I was a mental health specialist. I should also mention that this was before the American Psychiatric Association put DSM V (Diagnostic and Statistical Manuel) into effect. For these reasons, these cases are not a random or full sample of what we saw at the clinic, and the diagnostic nomenclature is used in a somewhat flexible and creative fashion. Many of these cases did not fit neatly into the usual categories, and often more than one diagnosis was involved. Reactions to Acute Traumatic Stress take many forms, depending on the person, his or her pre-existing medical and psychiatric conditions, and the particular family and the micro-circumstances of the trauma they experienced.

Guinee Clinic Cases

1. 11 year old Boy, with Arrested Pathological Grief Reaction, who was holding baby who was crushed to death in his arms, hallucinating dead baby's cries and screams of dying neighbors. Previous death of friend caused him to hallucinate voices 3 years early.
2. 41 year old Mother of 5, with acute **Anxiety Reaction**, with palpitations after daughter late getting home from school.

3. 46 year old Mother of 21 yo teacher son with **Prolonged Acute Mourning** when son was crushed at school, now bordering on depression. Serious insomnia, pre-existing high blood pressure.

4. 39 year old woman with **Acute Depression** who lost 1 child, an aunt, car crushed, business smashed, looted, severe insomnia, isolating herself, angry underneath. Given sleep meds, since anti-depressant costly.

Mirogane Clinic Cases

1. 35 year old woman with acute **Anxiety Reaction** due to multiple losses, continuing quake/after-quake shock reactions.

2. 41 year old man, a **Chronic Alcoholic,** drunk now for 8 years, with underlying **Acute Post-traumatic Stress Reaction.** Tear drops from his eye over lost family, word torrents. 'Devil drink' removed from sister by Voodoo priest, put in him.

3. 21 year old female medical student. **Acute Traumatic Stress Disorder**, after being trapped upside down, rescued after two days after listening to fellow students dying all around her.

4. 22 year olf girl, with hysterical **Psychophysiologic Reaction**, whose brother died, who develops Globus Hystericus,'the stifled cry syndrome' when she begins to remember and mourn.

5. 27 year old Man, with an unusual **Psychophiologic Reaction** in the form of a severe hysterical negative auditory hallucinations, beginning after hearing screams of injured, dying students and teachers, plus nightmares, insomnia. This could be under a definition of **Acute Post-tramatic Stress Disorder**.

6. A 5 year old **Developmentally delayed** little girl, with **seizures** who lost her medication and discovered her doctor was injured. her seizures were not well controlled we adjusted and gave her seizure medications.

COMMUNITY ANTHROPOLOGY, VOODOO, AND COLLABORATION

What is missing in this chapter so far, especially one written by a former anthropologist now working as a disaster psychiatrist? The answer is: a description of the Haitian social structure and how understanding it can be harnessed to heal a fractured, earthquake-torn country. Knowing just enough about social structure can facilitate the care and teaching of disaster psychiatry in any country.

Putting the rare Haitian 'Elite' aside, Haiti has a small urban middle class, often recently evolved from and still connected with their country peasant roots. This middle class group is urban, enterprising, and usually Catholic or Protestant. In Haiti, at least 85% of the population practice Vodun or Voodoo. Out of experience, necessity, and wisdom, the Catholic church is quite flexible, and accommodates Vodun. As a wise peasant said to me, with some exaggeration, most peasants practice Vodun, and 100% are Catholic. Among the small group that are Protestant, the Episcopals tolerate Vodun, while the Pentecostals are fundamentally opposed to it, feeling it is incarnate Devil worship.

Why 'incarnate'? Because virtually all who practice Vodun are possessed during ceremonies by Voodoo deities, and in that state give divine guidance and commands to the peasants, and even to the Vodun priest. After the possession state is over, the peasant has complete amnesia for what said or did. My Yale thesis is about Voodoo, and in particular, its religious form of ritualized multiple personality. While this may seem alien to most of us, Vodun is a rich, wise, and constructively prophetic religion from African roots. The Vodun priests are usually mature, influential leaders of the peasant community. The social structure, and often the placement of their Kay or small houses, center around the Houmfor, or Vodun temple. Vodun is a 'syncretic' religion, incorporating Catholic rituals and Saints in their ceremonies.

Why mention all this? Because the Vodun priest in many ways functions as political boss, advisor, doctor and psychiatrist for his people, and therefore Vodun supports and controls the vast majority of the Haiti population. If a hurricane or an earthquake kills a priest and destroys his houmfor, the center of social guidance and control are wiped out, leaving the peasants leaderless, intensifying their trauma and retarding their healing 'retribalization'. On the other hand, for any medical and mental health group trying to work with the peasants and multiply their healing effectiveness in enduring ways, they are well advised to form a respectful collaboration with their local Vodun priests and work with them. These priests, often sophisticated and smart about collaboration, will 'treat' disaster victims in their area, but also know their limit, and welcome having clinics to refer out to when a case is beyond their capacity. They like to get credit for being able to refer. And, when the medical clinic completes its acute care, the doctors can refer the peasant back to his local community and Vodun priest. Hospitals and local clinics are rare even if not collapsed, but Vodun priest are everywhere.

When I was a Yale anthropologist and a naïve premed, I practiced my brand of medicine in Haiti because peasants came to me for help. Once I made the mistake of trying to treat a burn case without realizing he was under the care of a Priest, and I got in big trouble. Later, he took me aside, explained how things work in Haiti, and we got our signals straight. When I apologized and shaped up, he became a great friend and collaborative guide for me. It took some work to convince the IMC doctors of the utility and wisdom of actually collaborating with Vodun priests. Most were middle class, and either Catholic or protestant. Some were still in touch with their peasant and religious roots, while others were uncomfortable or even alienated from their family background. Once the doctors and nurses thought it through, they became adept at utilizing their own internal cultural resources. 'Any port in a storm' I heard one doctor saying to a nurse. Though in truth they found it still much easier to collaborate with Catholic and Protestant institutions.

One evening, after a little medicinal Barbancout rhum, I decided to tell my version of Haitian history to explain why Voodoo was so important to the peasants. What follows opened a few medical eyes.

USEFULNESS OF HISTORICAL PERSPECTIVE

Haiti was once a lush tropical paradise called the Pearl of the Antilles, with beautiful forested mountains and rich alluvial plains--the crown jewel among France's prosperous colonies. Her plantations devoured vast numbers of slaves, many dying on slave ships even before reaching their harsh new world. To subdue them, French slave masters broke and scattered their families and tribes, rendering them totally dependent on their overseers.

But the slave masters made one fatal mistake. They allowed the slaves to hold religious ceremonies in the dead of night. From shared African roots, ancient and powerful Vodun Loa, their ancient Rada gods, cool and wise, appeared during ceremonies, biding their time. Through spirit possession, these Gods took over the consciousness of the downtrodden slaves, one at time during ceremonies, boldly speaking of ancestor worship, community and hope. Slowly, the new world religion of Voodoo, or Vodun, was born and spread among the plantations throughout Haiti. Led by powerful priests called Houngans, Voodoo became a healing and unifying force for the slaves during their darkest hours.

Even so, life on the plantations became more unbearable near the end of the 18th century. Suddenly, hot new Loa, Petro gods, began possessing slaves, demanding blood and revenge. Out of this Voodoo hotbed, a volcanic eruption shook Haiti, as these bloodthirsty gods and their angry Houngans ignited rebellion across the land. After thirteen years of guerilla war and 20,000 troops sent by Napoleon, Haiti's slaves stunned Europe with a military victory--becoming the only slave colony to win freedom, making Haiti the second country after the United States to gain independence in the Western Hemisphere.

From this glorious moment in 1804, Haiti has experienced a tragic decline. Her succession of governments, often self-serving or corrupt until the most recent, have allowed her people to plunge into abject malaria-ridden poverty. With virtually no roads or phones, no trains or power, Haiti's hearty but illiterate peasants barely survive. Stripping the mountains of trees for charcoal, the country now suffers dangerous erosion, flooding and mudslides. Thousands have been buried alive during tropical storms and hurricanes, and 150,000 died in this most recent earthquake. Though the plains are still fertile, overpopulation and excessive land division have left the peasants that remain eking out their existence. With few doctors and the highest infant mortality rate in the Western hemisphere, peasant life is so difficult that hundreds of boat people try to escape to the United States each year--abandoning their beloved island of blood and bougainvilleas. Drug traffickers, dealing especially in the white snow of cocaine, exploit Haiti. Against this seemingly hopeless backdrop, the Haitian peasants live with hearty enduring enthusiasm--their social fabric knit together by their predominant religion, Vodun, with their priests providing spiritual direction, compelling social organization, and medical-psychological consultation—just as in the days of the Haitian Revolution.

As if the earthquake were not enough of a challenge, shortly after returning to the US, as I was reviewing my field notes and diary, Hurricane Thomas roared through Haiti, followed by the Cholera epidemic brought to Haiti by an unwitting United Nations Nepalese caretaker soon taking a huge toll. Even so, Haitians are an amazing people, true survivors, and yet I worry there are limits to the suffering they can endure. I hope the world relief effort begins to take firmer hold very soon.

CONCLUSION

In conclusion, I hope this first-hand description of front line clinical work in Haiti illustrates many of the essential clinical and training principles of psychiatric trauma work and gives an immediate sense of the experience of care giving under dire circumstances. Perhaps the most surprising thing I learned during my work in Haiti is the fact that, despite the vast number of people traumatized, the diagnosis of Post-Traumatic Stress Disorder is a low-percentage outcome, perhaps around 13%. And by using the approaches utilized by the International Medical Corp, reflected in my lectures and clinical practice, that percentage can be drastically reduced. Prevention of chronic medical or psychiatric disease is perhaps one of the most important outcomes of this work. While Acute Traumatic Stress is common, early medical and psychiatric intervention can prevent that state moving on into chronic post-traumatic stress disorder. Given how resilient people are, especially the Haitian people, this approach really works.

At the same time, people who are vulnerable or have chronic pre-existing conditions come to these new clinics to receive help for the first time, or because they are cut off from their physicians or treating institutions. Providing care for them is expectable and essential, but follow-up is difficult because of the lack of facilities or local medical care. Recognizing this dilemma, even for the most hesitant or biased physician, leads in a practical sense to the strategy of using any surviving community leaders or healers. In Haiti, this means turning to Vodun *Houngans* (Priests), both for referrals and follow-up. They are the major resource in the community for healing, collaboration, and 'retribalization' (re-establishing community social structure. To a smaller degree, in other instances, one can turn to the Catholic, Episcopal or Pentecostal Churches. If medical groups try to 'go it alone' by trying to rely solely on doctor or hospital referral, they discover that resources are often scarce. Acute care is the major focus, per necessity, and rightly so. Acute care and prevention are the major goals. But community healing an essential reintegration are and enduring part of the solution.

REFERENCES

1. The Mw 7.0 Haiti Earthquake of January 12, 2010 USGS/EERI Advanced Reconnaissance Team Report V.1.1 February 23, 2010
2. Psychiatric Consultation to the Child with Acute Physical Trauma, in Annual Progress in Child Psychiatry and Child Development, Ed. By Stella Chess, M.D. and Alexander Thomas, M.D., Bruner-Mazel, New York, page 448-461, 1983

3. Different Faces of Trauma in a Three and a Four Year Old Girl, the International Institute of Object Relations Therapy, Summer Institute on Clinical Work Across Generations and Modalities, July 14-20, 2002

4. Haiti Fare Well (book) Kent Ravenscroft, MD 2011 Lulu.com ISBN 978-1-257-03187-0

5. Disaster Psychiatry in Haiti: Training Haitian Medical Professionals   (E-book) Kent Ravenscroft, MD 2013 International Psychotherapy Institute (E-Book Division)

6. Body Sharing: The Drug War, the CIA, and Haitian Voodoo by Kent Ravenscroft 2012 ISBN 978-0-557-29995-9 Lulu.com

7. The Divine Horseman: The Living Gods of Haiti (book)(film) Maya Deren Amazon.com

8. Secrets of Voodoo Milo Rigaud Amazon.com

9. Earthquake Psychiatric Relief Kent Ravenscroft, MD Images in Psychiatry, American Journal of Psychiatry, Sept 10, 2010

10. Experiences in Haiti Kent Ravenscroft, MD Washington Psychiatrist, Spring Issue 2013 Washington Psychiatric Society Bulletin

11. Helping in Haiti Kent Ravenscroft, MD IPI Bulletin, Vol. 13 Number 1 Spring 2010 International Psychotherapy Institute

12. My experience in Haiti: a brief report Kent Ravenscroft, MD World Psychiatry, volume 3, October, 2010

CITATION

Kent Ravenscroft (2015). Acute Psychiatric Trauma Intervention — The January 2010 Haiti Earthquake, Earthquake Engineering - From Engineering Seismology to Optimal Seismic Design of Engineering Structures, Prof. Abbas Moustafa (Ed.), ISBN: 978-953-51-2039-1, InTech, DOI: 10.5772/59867.

CHAPTER 7

Assessment of Seismic Hazard of Territory

V. B. Zaalishvili

Center of Geophysical Investigations of RAS, Russian Federation

INTRODUCTION

The new complex method of seismic hazard assessment that resulted in creation of the probabilistic maps of seismic microzonation is presented in this chapter. To study seismicity and analyze seismic hazard of the territory the following databases are formed: macroseismic, seismologic databases and the database of possible seismic source zones (or potential seismic sources - PSS) as well. Using modern methods (over-regional method of IPE RAS - Russia) and computer programs (SEISRisk-3 – USA) in GIS technologies there were designed some probabilistic maps of seismic hazard for the Republic North Ossetia-Alania in intensity units (MSK-64) at a scale of 1:200 000 with exceedance probability being of 1%, 2%, 5%, 10% for a period of 50 years, which corresponds to recurrence period of 5000, 2500, 1000, 500 years. Moreover, first the probabilistic maps of seismic hazard were made in acceleration units for the territory of Russia. The map of 5% probability is likely to be used for the large scale building, i.e. the major type of constructions, whereas the map of 2% probability should be used for high responsibility construction only. The approach based on physical mechanisms of the source is supposed to design the synthesized accelerograms generated using real seismic records interpretation.

For each of the zoning subject the probabilistic map of the seismic microzonation with location of different calculated intensity (7, 8, 9, 9*) zones is developed (the zones, composed by clay soils of fluid consistency, which can be characterized by liquefaction at quite strong influences, are marked by the index 9*). The maps in acceleration units show the similar results.

The complex approach based on the latest achievements in engineering seismology, can significantly increase the adequacy or foundation for assessments and reduce the inaccuracy in earthquake engineering and construction.

Realization of investigations on mapping of seismic hazard such as detailed seismic zoning (DSZ) based on the most advanced field research methods and analysis of every subject of the Northern Caucasus separately on a scale of 1:200 000 gives the possibility to merge a bit unavailable, at first glance, schemes into geologically and geophysically quite reasonable map of DSZ for the Northern Caucasus with equal scale system of the source zones.

ASSESSMENT OF SEISMIC HAZARD: GENERAL AND DETAILED SEISMIC ZONING

The seismic hazard of some territory represents a possible potential or a level of expected hazard, caused by geological structure features, tectonic movements, geophysical fields, macroseismical catalog, engineer-geological and hydrogeological structure etc. The adequate assessment of seismic hazard, at the same time is one of the important problems of engineer seismology. Unlike short-range and middle-range earthquake forecast, the involved assessment of seismic hazard, presented as seismic zoning maps, in fact is a long-ranged forecast of the earthquake strength and place.

One can mark out three types of analysis three consecutive stages of seismic zoning:

- general seismic zoning – GSZ or SZ, is realized in 1:5 000 000 or 1:2 500 000 scale
- detailed seismic zoning DSZ, was originally carried out for the most studied regions of perspective construction in 1:1 000 000, 1:500 000 scale or very rarely in 1: 200 000 scale.
- seismic microzonation – SMZ, in 1:25 000 scale or greater, contained in engineer investigation system.

The results of seismic zoning have to be the appropriate map creation GSZ, DSZ and SMZ. DSZ differs from GSZ in investigation scale. At the same time, in DSZ process may and must be studied all potential sources of possible earthquakes, which may be not taken in account, e.g. they have relatively small seismic potential during GSZ analyzing. It has to be mentioned, that in the real conditions the consequences of seismic hazard generation with that types of sources may have, if not great, but noticeably negative effect. At the same time both types of zoning are very similar, nothing to say about minuteness.

The third stage or stage of seismic hazard assessment in SMZ type has absolutely other physical meaning, in spite of similar name with GSZ and DSZ. The SMZ using allows to take into account the seismic properties of site soils, including physicomechanical and dynamical properties of soil.

The SMZ map traditionally is a normative part of Building Codes, and regularly is revised. At the same time during the map design only huge geology-geophysical zones are taken into account, which the seismicity determined.

The assessment of seismic hazard of the site is carried out using necessity and probabilistic methods. The probabilistic analysis of seismic hazard assessment includes alternative models of seismic sources, the earthquake returne periods, the seismic signal attenuation and distance dependence, and much vagueness, caused by careless information of some parameters, and by random character of seismic events. In the necessity analysis of seismic hazard assessment the vagueness is not considered, only the extreme seismic effect is estimated on the real site, using near earthquake sources with fixed magnitudes. There are many domestic and foreign algorithms and programs for this purpose.

Practically all the previous maps of seismic zoning, from the first map (1937) in the former USSR till the last but one map (1978) were necessity. They not take into account the main characteristic of seismic regime of seism active territory, although in the middle of 40th S.V. Medvedev (Medvedev, 1947) proposed to bring in seismic hazard zones internal differentiation including the strong earthquake return periods and assumed constructions durability. Then U.V. Riznichenko created algorithms and programs for seismic "shakeability" estimation (Riznichenko, 1966). But all these progressive development of domestic seismologists, like their other ideas were not brought in use. (Seismic zoning of USSR territory, 1980). At the same time these ideas were brought in use abroad, after analogous paper of Cornel K.A. (Cornell, 1968). And then western countries begun to create seismic zoning map in exceeding (or nonexceeding) probability of seismic hazard in given times intervals.

The vagueness conditions, are always presented in nature, so the necessity method in the seismic zoning is incompetent. The seismic zoning process must use only probabilistic methods. The risk is always presented, but it must be estimated and reduced to minimum. These ideas are presented in the new more progressive maps of Russia general seismic zoning - GSZ -97. For the first time in Russia was proposed to use the probability map kit GSZ -97 for different constructions (Ulomov, 1995). General map GSZ -97 is presented on fig. 1.

Wide spread usage of GSZ is caused by insufficient development of DSZ and distinct labor-intensiveness of its realization for researchers. Prof. Ulomov and his colleges use modern methods instead of ancient and out of date approach. In the same time the GSZ materials using sometime is impossible due impossibility to use more detailed information of regional and local materials including tectonical materials. The map generalization is enough for state overall planning, but is not enough for reliable estimation of real objects seismic conditions.

The process of Detailed seismic zonation is very complicated and expensive complex of geology tectonical, geophysical and seismical investigation for quantitative estimation of seismic effect in any site of perspective region (Aptikaev, 1986).

That type of investigation consists of all methods used in DSZ, but estimated quantitatively the source (background) seismic effects only on concerned site GSZ (more precisely for mean soil conditions or 2nd seismic category soils on site).

So, it is necessary to develop DSZ approach. The modern DSZ has clear and argumented content. There is huge Strong Motion Data Base with many records of soil velocity and acceleration, including South Caucasus Countries. Now, there are many modern computer programs, reliable digital velocity and acceleration registrators, now we may obtain many records of earthquakes. So, it is possible to realize DSZ purpose using reliable data. And, in spite of updating initial seismicity (UIS) for DSZ we have tye possibility to estimate site seismic hazard.

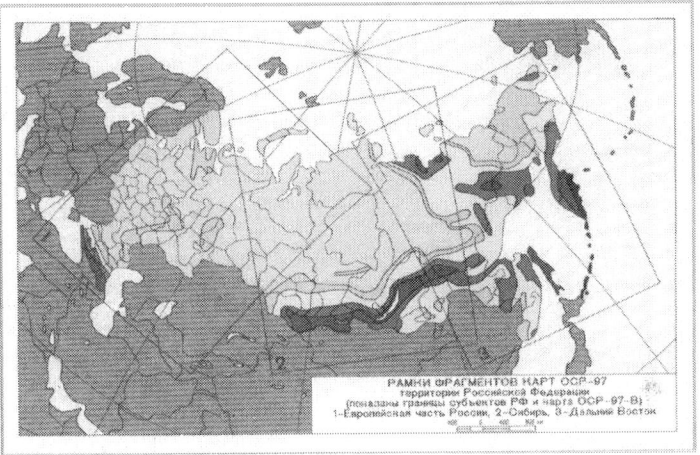

Figure 1: Map of General seismic zonation GSZ-97 of Russia.

It must be told, that UIS-DSZ methodic always formed parallel with GSZ methodic, but the scale differs, and some additional methods.

There are some methods that may be used in GSZ and DSZ for seismic generic structures (SGS) identifications, it is identification of zones of danger earthquake appearance (Nesmeianov, 2004).

Seismogeological Method

Using the first epicentral zone investigation in the end of XIX and beginning of XX centuries Abich G. and Lagorio A.E. find out the dependence between earthquake and tectonic structures. Mushketov I.V. writing about Vern earthquake 1887 year, told that Turkestan earthquake is connected with discontinuous disturbance (Mushketov, 1889). He wrote earthquakes "culminate on the boundary of the most huge and new disturbance" (Mushketov, 1891). Besides, he wrote that some groups of earthquakes are connected with lines, transversal to common stretch of rugosity, e.g. connected them with transversal structures in modern terminology. K.I. Bogdanovich analyzing Kebi earthquake (1911) consequences in the North Tien Shan, introduced new term – seismotectonic element, and for the first time proved the seismic shock migration inside seism active zone.

So, seismogeological method was able to connect strong earthquakes with tectonic structures. Those bonds later were named as geological seismicity criterion and were used in other methods.

Seismotectonical Methods

Seismotectonical method was introduced in the end of 40th years of XX century by Gubin I.E. when investigate Garm region on the Pamirs - Tien Shan border. He connected earthquakes with discontinuous zones some tenth km wide. He wrote, that "seismogenity degree" is stable all over the zone, "seismicity degree" may be ascribed to other similar zones, "if this structures, using geological data, are connected by mutual evolution process with equal intensity". This method (Gubin seismotectonics law) says, that in a given geological medium in the active structures of the same type and size, maximum earthquakes, originate from the rock displacement along the active rupture, have equal magnitudes and sources. Seismotectonical method accents on geological seismicity criterion – the velocity of young rupture displacement.

Seismostructural Method

Seismostructural method developed in the mid-50's, by V. Belousov, A. Goriachev, I. Kirillova, B. Petrushevsky, I.A. Rezanov, A.A. Sorsky, but most fully reflected in works of B. Petrushevsky. Earthquakes associated with large structural complexes-blocks allocated by using the historical-structural analysis and discontinuous joints.

Large-scale analysis blocks allowed to associate with them (and the underlying faults) varied range of depths of earthquakes (most profound on the articulation of the Pacific with Eurasian and the American continents). Picture of the strong earthquakes focuses with different three-dimensional structures of the Earth's crust was further developed in the works of G.P. Gorshkov (Gorshkov, 1984). However, this promising direction needs to be fleshed out.

Tektonophysical Method

Tektonophysical method developed in the second half of the 50-ies by M.V. Gzovsky. The method connects the earthquake with the maximum tangential stresses area, which is in conjunction with the maximum gradients of average speeds of tectonic movements and breaks. The energy of the earthquake was put by M.V. Gzovsky in dependence on a number of factors. But a precise calculation is impossible because the mechanical properties of the Earth's crust and its viscosity in Maxima tangential stresses can be evaluated only in qualitative terms.

Method of Allocating Quasihomogeneous Zones

In the late 50-ies started to be developed method of allocating quasihomogeneous zones of earthquakes for one or all geological and geophysical criteria, some of which have tectonical nature. However, these criteria have not been effective in a number of regions.

Since the choice of number and encoding parameters and their combinations are endless, equally infinite may be variants of map M_{max}. In connection with this were analysed practically all existing geologic-structural, seismic and geophysical maps for the territorial zoning using seismotectonic capacity (combined geological criteria reflecting the characteristics of the medium properties and the intensity of tectonic process), described in conventional units on a reference site. Based on mathematical patterns is forecasting of magnitude Mmax with reference site to the rest territory. Let's note the approach developed by Reisner and Ioganson, where reference sites were used in all of the zones of the planet. The analysis involved areas with variety of tectonic properties, where the seismicity criteria are mixed. Naturally, the common criteria were often not the fault criteria (thickness of the Earth crust, the heat flux density, height, isostatic gravity anomalies, the depth of the consolidated Foundation, etc.) The method later became known as the "extraregional method".

Method of Seismoactive Nodes

Structural refinement of earthquakes has allowed to V.M.Reiman at the turn of the 50 's and 60's to make an idea using the Central Asian material of disjunctive nodes in which the strong earthquakes concentrate or sejsmogenetic nodes. Later became actively used the term the seismically active sites. The best method was developed by E.Y. Rantsman (Rantsman, 1979), which extended the scheme to many orogenetic regions of the world. E.Y. Rantsman links with sites the earthquakes epicenters, stressing that "the earthquakes focuses can reach hundreds of miles away and go far beyond the morfostructural nodes". To classify the structures of seismicity was proposed the complex system of formalized criteria (distance from the edges of the site, type of terrain, maximum height, and area of friable deposits) and mathematical apparatus. Study of seismogenerating structures made it possible to include cross rises in the number of structures that make up the nodes (Nesmeyanov, Barkhatov, 1978).

Paleoseismological Method

Paleoseismological method (V.P. Solonenko, V.S. Khromovskikh, A. Nikonov, etc.) allows using paleoseismodislocations to trace possible seismic sources zones (PSS zones) and estimate their magnitude and seismic intensity. To evaluate these parameters the seismotectonic dislocation is used. Currently, there are many formulas (public and regional), describing the statistical associations between seismodislocations (length, amplitude displacement) and seismological parameters (magnitude, the depth of the epicenter, intensity of seismic vibrations) of earthquake. To determine the occurrence frequency of earthquakes it is necessary to have a reliable assessment of the age of paleo-seismic dislocations. All possible approaches are used: geological-geomorphological, archaeological, historical data and radiocarbon dating of sediments, broken by seismodislocation and later. Dendrohronological method is used, which takes into account the changes in the growth of trees associated with earthquakes, as well as lihenometrical method of dating seismogenical samples and dedicated to its some species of lichens.

Detailed Seismic Zoning

As an example, the assessment of seismic hazard, let's consider some estimations on the DSZ level in the territory of North Ossetia (Zaalishvili & Rogozhin, 2011).

On the basis of an analysis of methods for identification of PSS zones was elected out of regional seismotektonic method to objectively identify the sejsmogenic sources. Despite some shortcomings, the entire method is characterized by the quantitative indicators and has strong decision-making apparatus. The method has been used by prof. E.A. Rogozhin in North Ossetia when solving various scientific tasks. In addition, using this method some similar tasks were solved not only for Russian but also overseas territories (Israel, Italy etc.) (Rogozhin, 1997, 2007; Rogozhin et al., 2001; Rogozhin et al., 2008). At the same time, this does not preclude obtaining reliable results and other known methods.

The methodology used in most probabilistic seismic hazard analysis was first defined by Cornell and as usually accepted it consists of four steps (Reiter 1990, Kramer 1996): 1. Definition of earthquake source zones (SSZ), 2. Definition of recurrence characteristics for each source, 3. Estimation of earthquake effect and 4. Determination of hazard at the site. The probabilistic hazard maps for the territory of under study was compiled and we shell describe in brief this works according to the above noted steps.

Definition of Earthquake Sources

As a rule, today probabilistic assessment of seismic hazard is used all over the world for the identification of seismic loads for the engineering projects. The probabilistic approach is a more systematized method for the assessment of quantity, sizes and location of future earthquakes (Bazzuro & Cornell, 1999; Cornell, 1968; McGuire, 1995) than any other methods. Formal procedures for the probabilistic assessment include the determinations of spatio-temporal ambiguities for the expected (future) earthquakes. The computer program EQRISK of McGuire became the main stage in the method development (McGuire, 1976). The program became widespread and is very popular up to present day. In this connection the probabilistic assessment of seismic hazard is often called Cornell McGuire's method. The program includes integration on ambiguities distribution.

The Caucasian region is characterized by high intensity of dynamic geological processes (McClusky et al., 2000) and hazards, connected with them, of both natural and technogenic character. The most clearly expressed among these hazards is seismicity, which is accompanied with wide range of secondary processes. Earth surface ruptures, activation of known earlier inactive faults, landslip phenomenon, collapses, avalanches, creep and subsidence of the earth surface, activation of surface structures, soil liquefaction and other hazardous phenomena can be noted among them.

The investigations on determination and parameterization of the seismic source zones in recent decades has been realized by V.P.Solonenko, V.S.Khromovskikh, E.A.Rogozhin, V.I.Ulomov, V.G.Trifonov, I.P.Gamkrelidze and others (Gamkrelidze et al., 1998; Paleoseismology of Great Caucasus, 1979; Nechaev, 1998; Rogozhin et al., 2008; Trifonov, 1999; Ulomov et al., 1999).

On basis of the results of the active faults study located southward of the Great Caucasian ridge, parameters of seismic source zones were chosen according to data of I.P.Gamkrelidze work (Gamkrelidze et al., 1998) and to the north of the ridge they were chosen on data of E.A.Rogozhin and others (Rogozhin, 1997). According to the results of the executed expert evaluation of seismic potential (Mmax) the maps of seismic sources zoning of the territory of North Ossetia (zones of possible seismic sources - PSS zones) were made up.

A new original method of more accurate ascertainment of the boundaries of seismogenic source (fault) active part and assessment of the potential of seismic source hazard (at works of detailed seismic zoning – DSZ) has been worked out in recent years (Rogozhin et al., 2008).

Let's consider the process of territory seismic hazard assessment for explanation of procedure usage by the example of the Central Caucasus (the territory of the Republic of North Ossetia-Alania).

PSS zones are referred to the active fault systems, singled out on a basis of interpretation of the materials of remote sensing and geological data. Decoding of multispectral three-channel space images of Landsat–4/5 (resolution 30 m) and Landsat–7 (resolution 15 m) was realized. Decoding of space satellite photos was executed in colored multispectral variant as well as in black-and-white variant. Different variants of the image synthesis were used for the analysis of polyzonal scanner pictures. Besides, identification of the lineaments was also executed separately on channels. Combined deductive – inductive approach was used for lineaments identification: integrated structures were decoded on the base of strongly generalized images with the following zooming in for detailing and vice versa local peculiarities of tectonic and exogenous structures with the following zooming out and generalization. The method of stepwise generalization was used with quantization on the scale levels 1:25000; 1:50000; 1:100000; 1:200000; 1:300000; 1:400000; 1:500000. In the scale range 1:25000 - 1:1 500000 space photomap on basis of snapshots Landsat-7 is used and in the range 1:500000-1:2 millions – space photomap, created on basis of Landsat–4/5 snapshots.

Extensive lineaments systems were identified with known faults, which were qualified on modern stage as active. The name of PSS zones was formulated on basis of faults and large settlements names. Morpho-kinematics of active faults is the base for qualification of seismic displacements kinematics in PSS zones. Hypocenters depth of expected earthquakes was calculated from the depth of fault plans, the depth on geophysical anomalies data and from the magnitude of expected events.

Maximum magnitude of expected earthquakes (seismic potential, M_{max}) was assessed on the results of usage of the over-regional seismotectonic method of seismic hazard assessment, offered by G.I.Reisner. Usage of this method, foundation of which is described in the number of publications (Reisner & Ioganson, 1997; Rogozhin et al., 2001), showed that the Northern Caucasus is the region of very high seismic hazard.

In 2007 it was determined on data of field investigations that for the urbanized territories of North Ossetia the most hazardous are Vladikavkaz, Mozdok, Sunzha and Tersk PSS zones (table 1), (Fig.2)(Arakelyan et al., 2008; Rogozhin et al., 2008).Parameterization of seismic sources was made after creation of these maps, i.e. maximum possible magnitude M_{max} for each seismic source was assessed. This is the most difficult problem in the process of parameterization of PSS zones. M_{max} was determined on the data of a number of authors (Chelidze, 2003; Rogozhin, 2007).

The second essential parameter, which characterizes expected earthquakes, is sources depth range, where the majority of seismic events with corresponding magnitude generate. According to the numerous investigations, Caucasus is the region with upper crust part location of seismic sources – their depth doesn't exceed 20–25 km (deeper seismicity is observed in Tersk-Sunzha zone in the area of Grozniy city and in Caspian Sea). As sources distribution on depth for this region wasn't executed, average value of depth (equal to 10 km) was taken for calculations (see table 1).

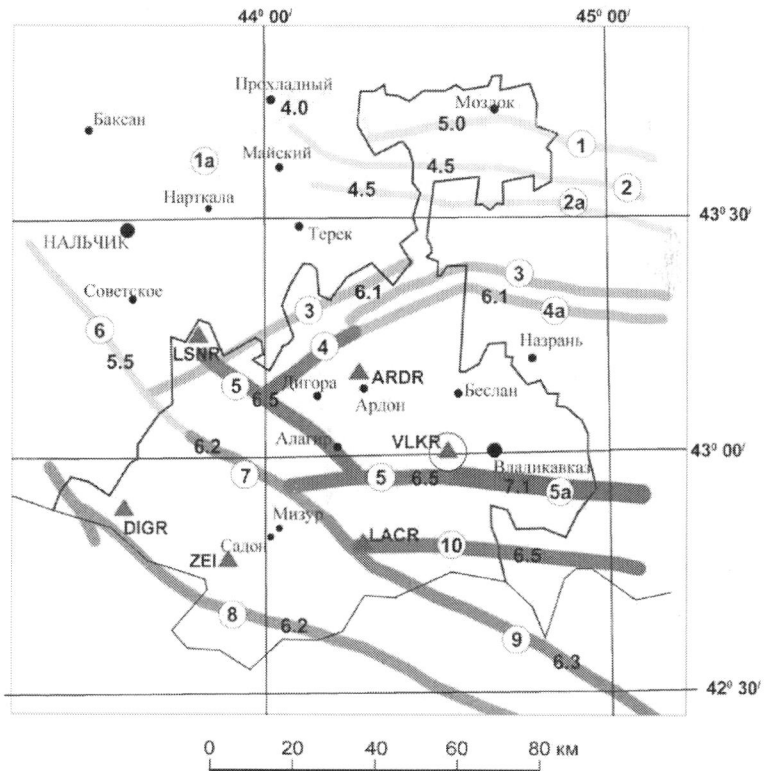

Figure 2: Map of PSS zones of the territory of the Republic North Ossetia-Alania (Rogozhin, 2007). Red triangles – basic seismic stations in the region. Blue and black lines are the state borders of North Ossetia.

Table 1: PSS zones for North Ossetia characteristics (numbers in the rings on Fig.1)

№	PSS zone	Magnitude	H, km	Kinematics.
1	Mozdok eastern	5.0	10	reverse faulting
1a	Mozdok western	4.0	5	strike-slip
2	Tersk	4.5	5	reverse faulting
3	Sunzha northern	6.1	15	reverse faulting
4	Sunzha southern (western branch)	6.5	15	strike-slip
4a	Sunzha southern (eastern branch)	6.1	15	reverse faulting
5	Vladikavkaz (western branch)	6.5	15	reverse faulting
5a	Vladikavkaz (eastern branch)	7.1	20	reverse faulting
6	Nalchik	5.5	10	strike-slip
7	Mizur	6.2	15	strike-slip
8	Main ridge	6.2	15	reverse faulting
9	Side ridge	6.3	15	reverse faulting
10	Karmadon	6.5	15	reverse faulting

Definition of Reoccurrence Characteristics

For the assessment of ratio parameters between reiterations during the process of execution of a number of investigations on the international projects the earthquake catalogue was checked and specified. The seismicity in each source zone was analyzed on basis of catalogue usage: New Catalogue... 1982, Corrected Catalogue of Caucasus, Institute of Geophysics Ac. Sci. Georgia (in data base of IG), the Special Catalogue of Earthquakes for GSHAP test area Caucasus (SCETAC), compiled in the frame of the Global Seismic Hazard Assessment Program (GSHAP), for the period 2000 BC - 1993, N.V. Kondorskaya (editor), (Ms>3.5) Earthquake catalogues of Northern Eurasia (for 1992-2000), Catalogue of NSSP Armenia, Special Catalogue for the Racha earthquake 1991 epicentral area (Inst. Geophysics, Georgia) and also the Catalogue of NORTH OSSETIA 2004–2006.

Corrected Catalogue of Caucasus contains data for more than 61000 of earthquakes, including 300 historical events (Byus, 1955a, 1955b, 1955c; New Catalogue of strong Earthquakes in the USSR..., 1982), which happened during 2000 years. This catalogue was checked and corrected. Some hypocentral parameters of earthquakes were recalculated.

Threshold of magnitude for the whole catalogue and a and b values of the frequency-magnitude law were determined for large tectonic zones, as their calculation for certain PSS zones was impossible because of data absence. Value of b of the frequency-magnitude law is determined by formula of Gutenberg-Richter:

$$\lg(N/T) = a - bM \qquad (1)$$

where a and b are parameters, the inclination and level of recurrence graph at M = 0.

For each PSS zone (both linear and square) frequency of earthquake origination was studied on basis of observed seismicity. For study of Gutenberg-Richter ratio earthquakes were referred to the separate faults or PSS zones taking into account accuracy in epicenter determination. Because of the shortage of data about accuracy of location determination average model was accepted. This model supposes that mistakes have normal distribution with standard deviation equal to 3–4 km. Distances from each event to the all PSS zones were measured and only zones, which were on closer distances from the event than three standard deviations, were taken into account. Based on distances value, weighting coefficient was assigned to each zone, from the curve of density distribution of the standard deviation possibility.

Estimation of Earthquake Effect

Earthquake effect was estimated using two different parameters: macroseismic intensity and peak ground acceleration (PGA). Macroseismic intensity (MSK scale) was traditionally used for seismic zonation in former USSR. Macroseismic and instrumental data on 43 significant earthquakes occurred in Caucasus were revised to obtain the necessary information (Javakhishvili et al. 1998). Data on 37 earthquakes was selected and in some cases were compiled new isoseismal maps in the 1:500 000 scale. In a process of computations was observed a fact that the value of the attenuation coefficient in vicinity (within the limits of the first three isoseismals) of the source of the Ms>6 earthquake is very high (~4.5-5.0), in comparison with small and moderate events (~3.4). This fact has been tested on the other Caucasian strong earthquakes (Ms>6) and in general has been confirmed. In spite of the lack of data in the first approximation the equation of correlation in this case obtains the following form for small earthquakes:

$$I = 1.5M_s - 3.4\lg(\Delta^2 + h^2)^{1/2} + 3.0 \qquad (2)$$

And

$$I = 1.5M_s - 4.7\lg(\Delta^2 + h^2)^{1/2} + 4.0 \qquad (3)$$

for large events.

The attenuation model according to the (2) formula is given on fig.3.

It should be noted, that for hazard estimation we have used the second relationship. Besides that we have restricted maximal value in epicentral area for M=7, (6.5) earthquakes with intensity 9, M=6 (5.5) earthquakes with intensity 8, etc. this was done to avoid very high intensities in epicentral area. The epicentral areas were estimated using relationships for earthquake source sizes given in (Ulomov 1999).

On the other hand strong motion instrumental data in Caucasus and adjacent regions allows us to use PGA and spectral acceleration attenuation law for seismic hazard analysis. Since the installation of the first digital strong-motion station in the Caucasus area 451 acceleration time histories from 269 earthquakes were recorded (Smit et al. 2000). Based on the acceleration time histories recorded between June 1990 and September 1998 with the permanent and temporary digital strong-motion network in the Caucasus and adjacent area, 84 corrected horizontal acceleration time histories and response spectra from 26 earthquakes with magnitudes between 4.0 and 7.1 were selected and compiled into a new dataset. All time histories were recorded at sites where the local geology is classified as "alluvium". Therefore the attenuation relations derived in this study are only valid for the prediction of the ground motion at "alluvium" sites.

The calculation of the correlation coefficients and the residual root mean square was performed with the well-known Joyner and Boore two step regression model. This method allows a de-coupling of the determination of the magnitude dependence from the determination of the distance dependence of the attenuation of ground motion. Using the larger horizontal component for spectra of the selected acceleration time histories, the values of coefficients were obtained for the coefficients at different frequencies. Because it is easy to obtain peak acceleration from corrected acceleration time histories, empirical attenuation models with peak ground acceleration as dependent parameter have always played an important role in different seismic hazard and earthquake engineering studies. The resulting equation for larger horizontal values of peak horizontal acceleration is:

$$\text{Log PHA} = 0.72 + 0.44\,M - \log R - 0.00231 + 0.28\,p, \qquad\qquad (4)$$

$R = (D2 + 4.52)^{1/2}$, where PHA is the peak horizontal acceleration in [cm/sec^2], M is the surface-wave magnitude and D is the hypocentral-distance in [km]. p is 0 for 50-percentile values and 1 for 84-percentile.

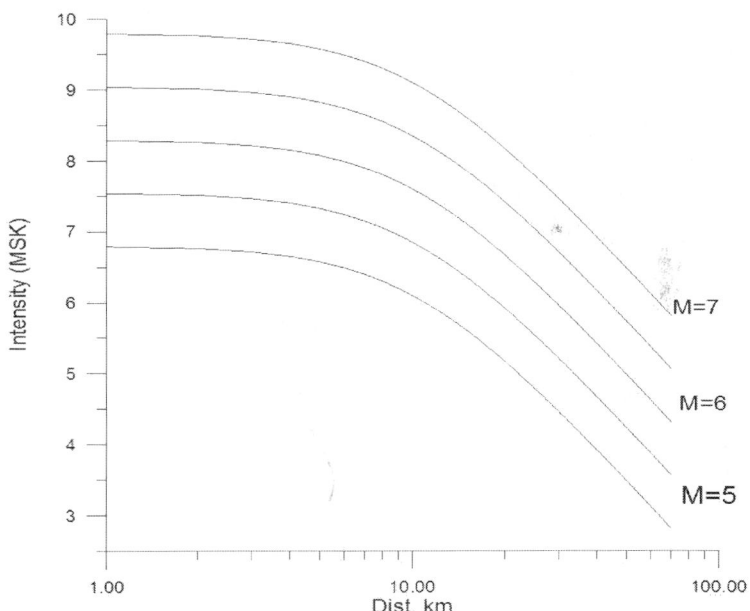

Figure 3: Attenuation model for intensity (MSK).

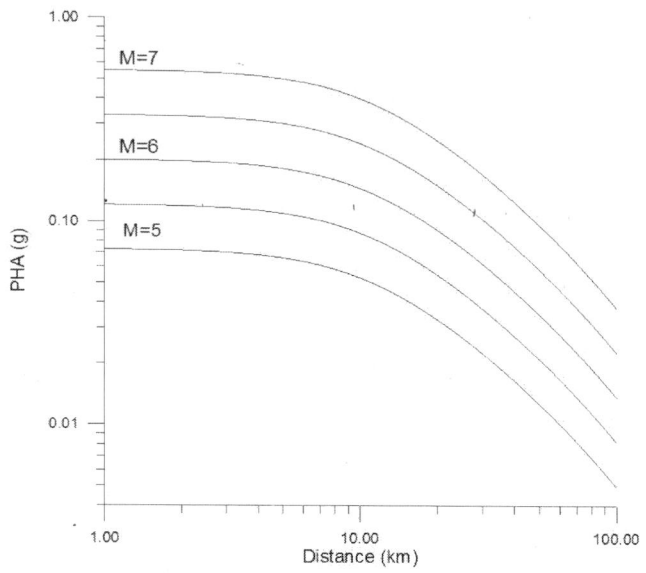

Figure 4: Attenuation model for acceleration.

It is important to bear in mind that all equations given above represent a best fit of the selected dataset, and therefore represent mean values about which there is a considerable scatter. In the case of the attenuation model for the larger horizontal value of the peak horizontal acceleration the predicted mean plus one standard deviation is equal to 1.91 times the mean value. The scatter of the pha-models is the same as similar models for Europe and Western North-America (Smit et al., 2000). The attenuation is shown on fig. 4.

The comparison of the attenuation relationships for peak horizontal acceleration with similar relations for other areas shows a good agreement with the models from Western North-America. It is obvious, that the attenuation in Europe is lower compared to the Caucasus and adjacent area. The predicted peak values in the near-field are higher than the corresponding values obtained with other European models (Smit et al., 2000).

Determination of Hazard

The probabilistic seismic hazard maps (the maps of detailed seismic zoning) have been constructed for the total area of North Ossetia in scale 1:200000 with exceedance probability for a period of 50 years (standard time of building or construction durability!) with 1%, 2%, 5%, 10% in GIS technologies, which corresponds to reoccurance of maximum probable earthquake for a period of 5000, 2500, 1000 and 500 years (Fig.5). The longer the period of time the higher the level of possible intensity. For a period of 500 years only a small part will be occupied by the zone of 7 intensity earthquake, for a period of 1000 years – 8 intensity and at 2500 years 9 intensity earthquake appearance, correspondingly.

Cornell approach, namely computer program SEISRisk- 3, developed in 1987 by Bender and Perkins (Bender & Perkins, 1987) was used for the calculations. The map of observed maximum intensity was compared with the maps of different periods of exposition and the most real map was chosen on a basis of the analysis of differences between the observed and calculated maps. According to these criteria the map of 5% probability with exceedance probability of 50 years can be recommended for seismic zoning of the territory of North Ossetia. Besides, for the first time probability maps of seismic hazard for Russian territory were made in acceleration units in scale 1:200 000 with exceedance probability for a period of 50 years - 1%, 2%, 5%, 10%.According to the Musson (Musson, 1999) conception, it is necessary to use the data, which is maximum approximate to the real engineering-geological conditions, at assessments of territory seismic hazard. For the territory of North Ossetia the exposition equal to 1000 years is the most approximate to real conditions for mass building. It is necessary to consider greater exposition, for example, 2500 years etc. for unique buildings and constructions.

The maps of 5% probability are likely to be used for the large scale building, i.e. the major type of constructions, whereas the maps of 2% probability should be used for high responsibility construction only (Fig.5).

One can see great hazard in the south of North Ossetia on the map, where exists the increased level of seismic hazard (due to powerful Vladikavkaz fault, lying nearby).

As a matter of principle it is possible to make maps in scale 1:100 000 etc., but it actually makes no practical sense. Although accuracy of such maps must be higher, adequacy of the results can be considered as doubtful due to absence of reliable data on local peculiarities of past, i.e. historical earthquakes display. Laboriousness (irretrievable) at that increases multiply.

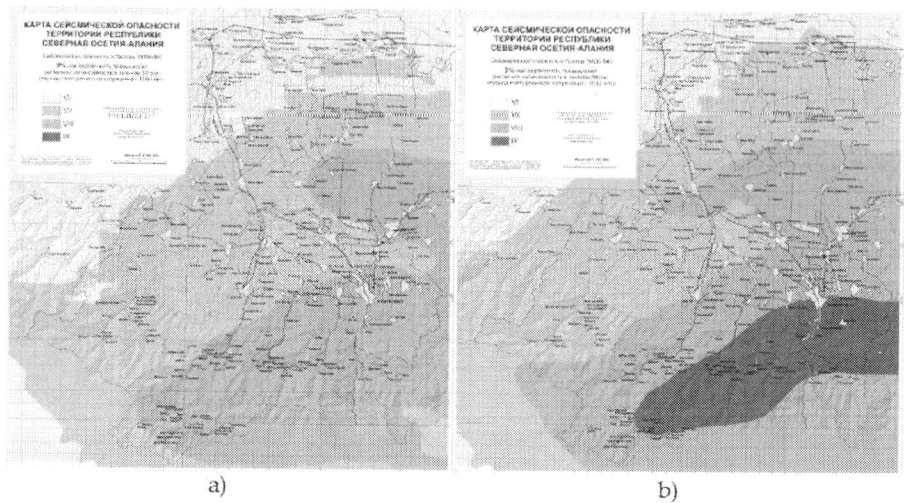

Figure 5: Probabilistic maps of seismic hazard (DSZ) in the intensities (MSK-64) with the exceedance probability 5% (a) и 2% (b) for North Ossetia territory and adjacent areas (Zaalishvili, 2006).

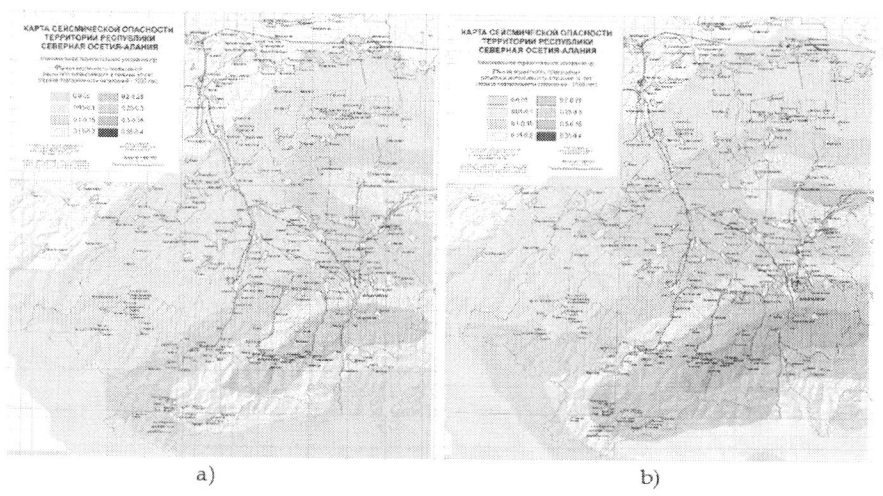

Figure 6: Probabilistic map of seismic hazard (DSZ) in accelerations (PGA)with exceedance probability 5% (a) and 2% (b) for North Ossetia territory (Zaalishvili, 2006).

The scientists from Vladikavkaz in collaboration with the colleagues from the Institute of Physics of the Earth of RAS not only offer to use large-scale maps but also decided to continue investigations and cover the whole Northern Caucasus in scale 1:200 000. So, maps of seismic hazard can be made up in scale 1:200 000 for the Republics of Chechnya, Ingushetia, Kabardino-Balkaria, Stavropol and Krasnodar areas and the other territories (Fig. 7). Taking into account, that faults and other peculiarities of the territory exist out of any boundaries, including state boundaries, it is possible to make unusual but quite physically proved single general map of detailed seismic zoning of the territory of Northern Caucasus in scale 1:200000, moreover, one can make them for different exposition times and accordingly for different probabilities. So, created maps of detailed seismic zoning of North Ossetia conform to earthquake realization once in 500 years, 5% - in 1000 years and 2% - in 2500 years. The level of seismic hazard grows with the time increase etc.

It is possible to make detailed maps of seismic hazard for the whole Caucasus, including Azerbaijan, Armenia and Georgia, due to the features of spreading of hazardous seismic sources, which «neglect» states' boundaries. It is also possible to develop the maps jointly with Turkey and Iran and it's real to include such countries as Israel, Egypt, and Lebanon etc.

Maps of detailed seismic zoning can be called «long-term» prediction maps. It means that long-term prediction of hazardous phenomena is realized on their basis and, correspondingly the place of earthquake-proof building-stock is determined

Essentially, the long-term maps of expected intensities locations are that of described maps of detailed seismic zoning. Indeed, that evacuate people from the hazardous territory before expected earthquake is impossible, but it is real to prevent population burring under destroyed or, to be more precise, differently damaged buildings, which is formed on basis of such maps. The more educated society is the less seismic risk, i.e. economic and social losses. So, the priorities are clear.

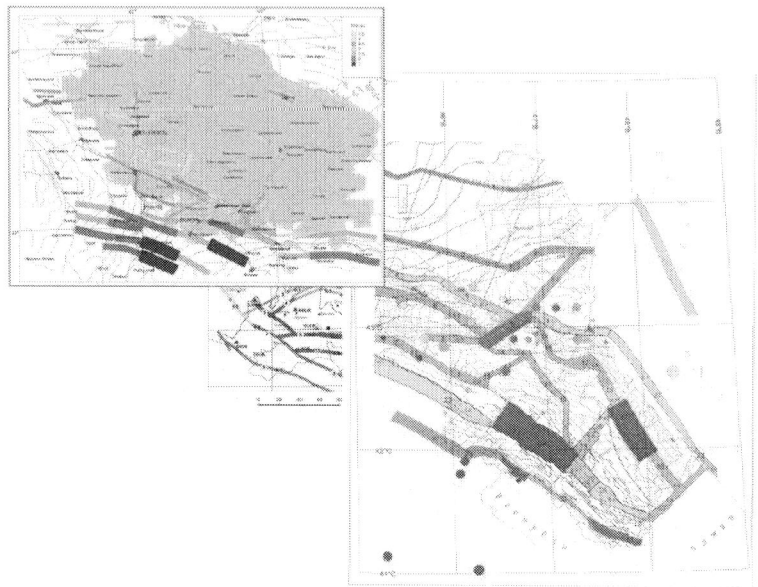

Figure 7: The mosaic of maps of hazardous potential seismic sources on the territories of the Northern Caucasus (model of the future joint map).

On basis of the given maps it is necessary to make up the maps of seismic microzonation (SMZ) of cities and large settlements of each certain subject of the Russian Federation with the usage of the most modern standard methods and tools, but in scale 1:10 000. The probabilistic maps of SMZ were first developed in the Center of Geophysical Investigations of Vladikavkaz Scientific Center RAS and RNO-A. Such maps of SMZ are direct and reliable base of earthquake-proof design and object construction.

Besides, it is necessary to note that at usage of the traditional units of macroseismic intensity the boundaries between different zones are characterized by sharp changes, which obviously do not correspond to the real situation of monotonous change of intensity for homogenous soil conditions of the investigated territory. No doubt, it will form evident inaccuracies at the assessment of the level of seismic hazard of this or that territory. The practical usage of artificial intensity subdivision, for example, in the form of 7.2 or 8.3 points is not validated enough from the theoretical point of view. So, firstly, it is not usually explained how these fractional assessments are obtained and, secondly, the following transition to the acceleration units (obviously, according to foreign data, as there are no acceleration records for forming reliable correlation in Russia), undoubtedly, forms considerable inaccuracy and it is hardly ever physically proved because of the formality of the parameter of «intensity» itself.

On the other hand, at seismic influence assessment at earthquake - proof design engineers use the acceleration values, (strictly speaking, conveniently) corresponding to specified intensities. Thus, it's assumed that design acceleration a = 0.1 g corresponds to the intensity 7 earthquake, 0.2g – to the intensity 8, 0.4g – to the intensity 9 etc. At the same time, network of digital stations dislocated on the Southern Caucasus installed in source zones of Spitak (Armenia, 1988), Racha (Georgia, 1991), Barisakho (Georgia, 1992), Baku (Azerbaijan, 2000), Gouban (Georgia, 1991), Tbilisi (Georgia, 2002) and other earthquakes collected seismic records for formation of database of accelerations for Caucasus. Namely it makes possible to design maps of the seismic hazard independently in units of PGA. Such maps for the territory of North Ossetia for exposition of 50 years with exceedance probability 1%, 2%, 5%, 10% in scale 1:200 000 were created (Fig. 6). It is obvious that at changing of smoothering step it is possible to obtain smooth variations of accelerations directly used as design impacts.

In contrast to the maps of general seismic zoning (GSZ) with a scale of M 1: 8000000 and, at the best, with the scale M 1:2500000 obtained maps of both types on a scale 1:200000 can be referred to the DSZ type maps.

Thus, these materials allow assessing seismic hazard on a detailed level, according to the known formulas to calculate the macroseismic field of seismic effects on a scale that may provide a reliable basis for SMZ.

SEISMIC MICROZONATION OF TERRITORY

Seismic microzonation (SMZ) actually is final stage of seismic hazard assessment. SMZ results are direct foundation for earthquake-proof construction. In the process of seismic microzonation sites with etalon ground conditions corresponding to specified seismic hazard level are specified. In Russia grounds with mean seismic properties for given territory are traditionally referred as etalon ground conditions. Usually these are soils with shear wave velocity of 250–700 m/s [SP 14.13330.2011]. In Georgia, for example, in dependence of specific engineering-geological situation etalon grounds in their seismic properties can be worst or mean for given territory. In USA firm rock grounds are referred as etalon. Seismic microzonation consists in intensity increments calculation caused by differences in ground conditions. Works on seismic microzonation are realized by instrumental and calculational methods.

Instrumental Method of Seismic Microzonation

Instrumental method is the main SMZ method. Exactly it urges to solve a problem of forming earthquake intensity forecast. At the same time the calculation method, which allows to model any definite conditions of area and influence features, is often characterized by more reliability. It has great importance to soil thickness with high power. Combined usage of both methods significantly increases results validity.

Seismic Microzonation on Basis of Strong Earthquakes Instrumental Records

It is supposed at usage of strong earthquakes records for SMZ purposes, that at some strong seismic influence the observing soil behavior is adequate to the display of their potential seismic hazard at future strong earthquakes (Nikolaev, 1965). This fact was the reason of stimulation of a number of large international scientific-research projects on organization of long-term instrumental observations with the help of powerful measurement systems in the Earth's different regions with high seismic activity for the purpose of obtaining the strong movements of soils, which are the base of buildings and constructions (the groups SMART-1 and SMART-2 on the Taiwan island etc.).

At the same time, presence of unit record of a real strong seismic influence at its inestimable value for SMZ often can't give the adequate forecast of soil behavior at a next following strong earthquake. This problem can be solved by creation of a number of records of seismic influences, generated by hazardous for the zoned territory active fractures, i.e. by zones of possible earthquake source (PES).

Seismic Microzanation with the Help of Weak Earthquakes Records

In the connection of the fact that strong earthquakes occur seldom, the intensity increments, as a rule, are assessed by records of weak earthquakes, when a linear dependence between the dynamic stress and the deformation takes place.

Soil conditions considerably change (fig. 8) the right shape of the original undistorted signal, incident from the crystal foundation. Complex shapes of isoseisms pointed out to the undoubtful link between the earthquake display intensity and soil conditions (Reiter, 1991).

Increase of the soil thickness depth (alluvium) considerably changes the character of earthquake records (Reiter, 1991) in the process of approaching the city (fig. 8).

Calculation of intensity increment with the help of weak earthquakes is realized by the formula (Medvedev, 1962; Recommendations on SMZ, 1974, 1985):

$$\Delta I = 3{,}3 \lg A_i / A_0, \tag{5}$$

where Ai, A0 are the amplitudes of investigated and etalon soils vibrations.

The usage of tool in the form of registration of strong and weak earthquakes needs the organization of instrumental observations in a waiting mode.

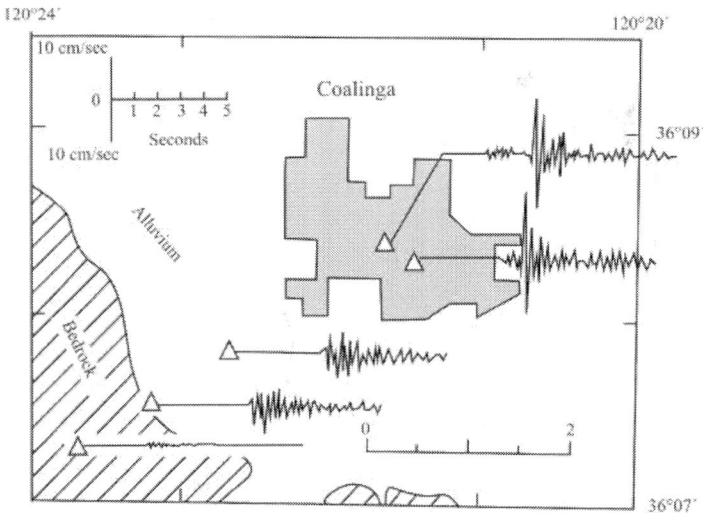

Figure 8: Scheme of California earthquake in Koaling sity.

Seismic Microzanation with the Help of Weak Earthquakes Records

In the connection of the fact that strong earthquakes occur seldom, the intensity increments, as a rule, are assessed by records of weak earthquakes, when a linear dependence between the dynamic stress and the deformation takes place.

Calculation of intensity increment with the help of weak earthquakes is realized by the formula (Medvedev, 1962, Recommendations on SMZ, 1974, 1985):

$$\Delta I = 3,3 \lg A_i / A_0, \tag{6}$$

where Ai,A0are the amplitudes of investigated and etalon soils vibrations.

The usage of tool in the form of registration of strong and weak earthquakes needs the organization of instrumental observations in a waiting mode.

Seismic Microzonation Using Microseisms

The results of microseisms observations (Kanai, 1952) are used as subsidiary instrumental tool of SMZ. Predominant periods are determined at that in order to assess resonance properties of soils and amplitude level of microvibrations. Strictly speaking, the reference of microseism on their origin to the purely natural phenomena is not quite correct. Numerous artificial sources, influence degree of which can't be controlled, undoubtedly, take part in their forming along with the natural sources (fig. 8.6).

Mpossibility of the compliance of necessary standard conditions of microseism registration and large spread in values of maximum amplitudes limit the usage of microseism for calculation of soil intensity increment. The above mentioned causes the application of microseism tool only in complex with other instrumental tools.

Spectral features for different sites are estimated by means of H/V-rations (Nakamura, 1989).

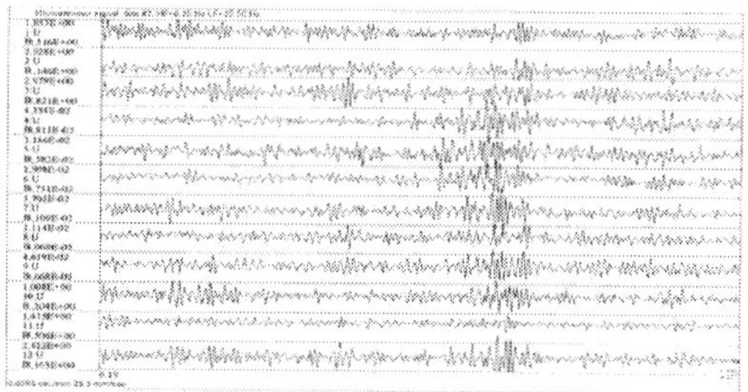

Figure 9: Microseisms records (10.07.1996, Voronezh Region, Russia).

Seismic Microzonation Using Explosive Impact

The intensity increment ΔI of the soils of the zoned territory is calculated by the formula (Medvedev, 1962; Recommendations on SMZ, 1974, 1985) at usage of weaker explosions:

$$\Delta I = 3,3 \lg A_i \,/\, A_0, \tag{8}$$

where A_i, A_0 are vibrational amplitudes of the investigated and etalon soils.

Execution of powerful explosions on the territory of cities, settlements or near the responsible buildings is connected with large and often insurmountable obstacles (technical and ecological problems, safety problems, labouriousness and economical expediency) and practically isn't used nowadays. This leads to the wide spreading of nonexplosive vibration sources.

Seismic Microzonation Using Nonexplosive Impulse Impact

The features of SMZ methods development led to the situation when the tool of elastic wave excitation with the help of low-powered sources (for example, hammer impact with m = 8–10 kilograms) has become the most wide spread in the CIS countries, in order to determine S- and P-wave propagation velocities in soils of the typical areas of territory. Velocity values are used in order to calculate the intensity increment using the tool of seismic rigidities by S.V.Medvedev (Medvedev, 1962; Recommendations on SMZ, 1974, 1985):

Where $\rho_0 v_0$ and $\rho_i v_i$ is the product of the soil consistency and P-wave (S-wave) velocity – seismic rigidities of the etalon and the investigated soil accordingly.

The intensity increment, caused by soil watering, is calculated by the formula

$$\Delta I = K e^{-0,04 h_{GL}^2} \tag{9}$$

where K = 1 for clay and sandy soils; K = 0,5 for large-fragmental soils (with sandy-argillaceous filler not less than 30%) and strongly weathered rocks; K = 0 for large-fragmental firm soils consisting of magmatic rocks (with sandy-argillaceous filler up to 30%) and weakly weathered rocks; h_{GL} is the groundwater level.

The simplicity and immediacy of practical application of S.V.Medvedevs' tool, which is called the tool of the "intensities", led to its widespread in CIS countries and countries of Eastern Europe, Italy, USA, India, and Chile in 1970-es. The tool of the "intensities" was advantageously different from other tools by the immediacy, simplicity in initial data obtaining and its processing and independence from seismic regime of the territory. It to a certain extent hampered the development and making up of new tools. Unfortunately, the calculation results of predicted values of intensity increment are often quite incorrect as data of macroseismic observations of destructive earthquake consequences shows (Shteinberg, 1964, 1965, 1967; Poceski, 1969; Stoykovic and Mihailov, 1973).

By means of the special investigations it was determined that the reliability of calculated intensity increments considerably increases at usage of modern powerful impulsive energy sources (fig. 9).

The lowering of final results quality is to a certain extent caused by the fact that in the tool of "intensities" the seismic effect dependence in soils on frequency or "frequency discrimination" of soils (Shteinberg, 1965) and also the origin of typical "nonlinear effects" at strong movements isn't taken into account. A.B.Maksimov tried to remedy this deficiency by developing the tool, where frequency peculiarities of soils were taken into account (Maksimov, 1969):

$$\Delta I = 0.8 \lg \rho_0 V_0 f_0^2 / \rho_i V_i f_i^2$$

(10)

where f_0, f_i are predominant frequencies of etalon and investigated soils.

A.B.Maksimovs' tool didn't find wide distribution, as frequency differences of soil vibrations with sharply different strength properties (at usage of traditional for the seismic exploration of small depths low-powered sources) were insignificant and the calculation results on the formulas (9) and (11) were practically similar (Zaalishvili, 1986).

Intensity increment was determined by the following formula (Zaalishvili, 1986):

$$\Delta I = 0.8 \lg \rho_0 V_0 f_{\text{wa0}}^2 / \rho_i V_i f_{\text{wai}}^2$$

(11)

where f_{wa0}, f_{wai} are weight-average vibration frequencies of etalon and investigated soils.

Weight-average vibration frequency of soils was calculated at that on the formula [Zaalishvili, 1986]:

$$f_{\text{CB}} = \sum A_i f_i / \sum A_i$$

(12)

where A_i and f_i are the amplitude and the corresponding frequency of vibration spectrum.

Figure 10: Surficial gasodinamical pulse source (SI-32).

Seismic Microzonation Using Vibration Impact

At usage of a vibration source (fig. 10) the calculation of intensity increment is realized with the help of the formula (Zaalishvili, 1986):

$$\Delta I = 2\lg S_i / S_0,$$

(13)

where S_i and S_0 are the squares of vibration spectra of investigated and etalon soils.

The developed tool was used at SMZ of the territories of cities Tbilisi, Kutaisi, Tkibuli, single areas of the Bolshoy Sochi city. The tools' feature consists in the fact that it allows to assess soil seismic hazard without any preliminary investigations: at realization of direct measurements of soil thickness response on standard (vibration or impulse) influence. Later the formula was successfully used at SMZ of the sites of Novovoronezh Nuclear power-plant (NPP) with the help of an impulsive source (Zaalishvili, 2009).

Figure 11: Vibration source (SV-10/100).

Seismic Microzonation on Basis of Taking Into Account Soil Nonlinear Properties

The comparison of the absorption and nonlinearity indices with the corresponding spectra of soil vibrations shows that at higher absorption the spectrum square prevails in LF field and at high nonlinearity it prevails in HF field of the spectrum. In other words, the presence of absorption is displayed in additional spreading of LF spectrum region, and the presence of nonlinearity – in spreading of HF range.

All the mentioned allowed to obtain the formula for calculation of intensity increment on basis of taking into account nonlinear – elastic soil behavior or elastic nonlinearity (at usage of vibration source) [Zaalishvili, 1996]: $\Delta I = 3 \lg A_i f_{wai} / A_0 f_{wa0,}$

where $A_i f_{wai}$, $A_0 f_{wa0}$ is the product of spectrum amplitude on weight-average vibration frequency of investigated and etalon soils.

The formula (14) characterizes soil nonlinear–elastic behavior at the absence of absorption.

If the impulsive source is used at SMZ than the formula will have the form (Zaalishvili, 2009):

$$\Delta I = 2 \lg A_i f_{wai} / A_0 f_{wa0.}$$

Seismic Microzonation on Basis of Taking into Account Soil Inelastic Properties

As soil liquefaction and uneven settlement of the constructions are observed at strong earthquakes (Niigata, 1966; Kobe, 1995), the most actual problem of SMZ is to assess possible soil nonelasticity adequately and physically proved at intensive seismic influences.

In order to assess directly nonelasticity of soil, the special scheme of the realization of experimental investigations (fig. 11, a) with gas-dynamic impulsive source GSK-6M (with two oscillators) was used. Selected location of the longitudinal profile allowed to influence alternately by two emitters from adjoining and somewhat far radiation zones. In the spectrum of soil vibrations, caused by near emitter, the HF component, which quickly attenuates with distance (fig. 11, b), predominates. In case of influence by distant emitter to the soil surface, the LF component predominates in the spectrum of vibrations (fig. 11, c). In other words, at nonlinear-elastic deformations the main energy is concentrated in the HF range of spectrum and at nonelastic – in the LF range. The signal spectrum has the symmetrical form in the far and practically linear-elastic zone.

Elastic linear and nonlinear vibrations are characterized for the given source by the constancy of the real spectrum square, which is the index of definite source energy value, absorbed by soil (which is deformed by the source). The analysis of strong and destructive earthquake records and also the analysis of specially carried out experimental influences showed that at nonelastic phenomena spectra square of corresponding soil vibrations is not the constant value. It can decrease and the more it decreases, the less the soil solidity and the greater the influence value (Zaalishvili, 2009).

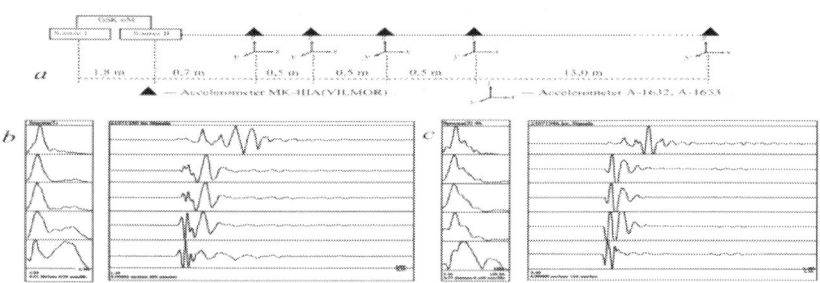

Figure 12: Investugation of site spectral features by means of GSK-6M seismic source: a) experiment scheme; b) record of second source impact; c) record of first source impact.

At usage of vibratory energy source, the whole number of new formulas (Zaalishvili, 2009) in order to assess soil seismic hazard with taking into account the values of their nonelasticity were obtained:

$$\Delta I = 2,4 \lg \left[(S_{ri})_n (S_{r0})_d / (S_{ri})_d (S_{r0})_n \right].$$

where $(S_{ri})_{n,d}$ and $(S_{r0})_{n,d}$ are the squares of real spectra of investigated and etalon soils in near and distant zones of the source.

$$\Delta I = 3,3 \lg (A_i f_{awi})_n (A_0 f_{aw0})_d / (A_i f_{awi})_d (A_0 f_{aw0})_n,$$

where $(A_i f_{awi})_{n,d}$ and $(A_0 f_{aw0})_{n,d}$ are the amplitudes and weight-average frequencies of investigated and etalon soils in near and distant zones of the source.

In case of powerful impulsive source usage the offered formulas will have a form:

$$\Delta I = 1.2 \left[\lg (S_{ri})_n (S_{r0})_d / (S_{ri})_d (S_{r0})_n \right].$$

where $(S_{Pi})_{бд}$ and $(S_{P0})_{бд}$ are the squares of real spectra of investigated and etalon soils in near and distant zones of the source;

$$\Delta I = 2 \lg \left[(A_i f_{awi})_n (A_0 f_{aw0})_d / (A_i f_{awi})_d (A_0 f_{aw0})_n \right].$$

where $(A_i\ f_{awi})_{n,d}$ and $(A_0\ f_{aw0})_{n,d}$ are the amplitudes and weight-average frequencies of investigated and etalon soils in near and distant zones of the source.

The formulas (17) and (18) are true only for loose dispersal soils. The formulas (17) and (18) were used at SMZ of the territory of Kutaisi city. Besides, with the help of the formulas (19) and (20) nonelastic deformation properties of soils in full-scale conditions on the site of Novovoronezh NPP-2 were defined more exactly (Zaalishvili, 2009). The formulas were obtained on basis of physical principle, which underlies the scheme, applied at the assessment of soil looseness measure (Zaalishvili, 1996,Nikolaev, 1987).

Calculational Method of Seismic Microzonation

Calculational method of SMZ is used in order to analyse features of soil behavior with introduction of definite engineering–geological structure characteristics of investigated site as initial data: values of transverse wave velocities, index of extinction, modulus of elasticity, power of soil layers, their consistency etc. Calculational method includes thin-layer medium, multiple-reflected waves, finite-difference method, finite-elements analysis (FEA) and other techniques.

One can take nonlinear soil properties into account in the problems of earthquake engineering by means of instrumental and calculation methods. The instrumental method of SMZ is the main method. Nevertheless it is quite often necessary to solve such problems using calculational method, which allows to model practically any conditions, which are observed in the nature. At the same time the practice reqirements lead to the necessity of calculation of soil vibrations for the conditions of their nonlinear-elastic and nonelastic deformations. At the solution of such problem it is assumed that elastic half-space behaves as linear-elastic medium and the covering soil displays strong nonlinear properties at intensive seismic or dynamic influences (Bonnet & Heitz, 1994).

Instrumental stress-sstrain dependences can be used, for example one obtained for plastic clay soil shown in fig. 12. The conception of the so-called soil bimodularity, offered by A.V.Nikolaev (Nikolaev, 1987,

Zaalishvili, 1996; 2000) is taken into account in the given dependence. Considerable differences in behavior of "weak" soils at compression and dilatation lie in the base of the phenomenon. Such soil is characterized at dilatation by quite small shear modulus.

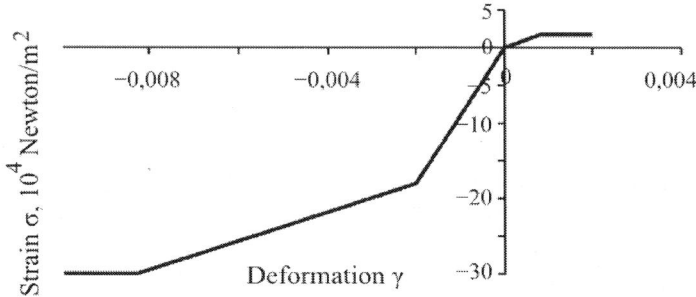

Figure 13: Instrumental stress-sstrain curve, showing property of soil bimodularity.

The solution of the given nonlinear problem for soils in the analytic form is based, as a rule, on considerable assumptions due to the complication of adequate taking into account behavior features of such complex system as the soil (Bonnet & Heitz, 1994). Therefore the numerical solution of nonlinear problems on the modern stage of knowledge is the most proved if the data of field or laboratory investigations is taken into account in these or those connections. Thus, the correlations, which are determined by the experimental investigations, are the basis of the solution of calculation nonlinear problems. In other words calculation programs for the solution of calculation nonlinear problems essentially are analytical-empirical. The most adequate programs are exactly like these (SHAKE, NERA etc.).

Equivalent Linear Model. Shake and Eera Programs

Equivalent linear model is one of the first models, which take nonlinear soil behavior into account. Equivalent linear approximation consists in modification of the model of Kelvin–Voight (for taking some types of nonlinearity into account) and, for example, is realized in the programs SHAKE (Schnabel et al., 1972) and EERA (Bardet et al., 2000).

Equivalent linear model is based on the hypothesis that shear modulus G and attenuation coefficient ξ are the functions of shearing strain γ (fig.

18.1). In the programs SHAKE and EERA (Equivalent-linear Earthquake site Response Analyses) the values of shear modulus G and attenuation coefficient ξ are determined (in the process of iteration) so that they correspond to the deformation levels in each layer.

IM Model. NERA Program

In 2001 realization principle, which was used in the program EERA, was applied in the programming of NERA (Nonlinear Site Response Analysis) (Bardet, Tobita, 2001), which allows to compute soil thickness nonlinear reaction on seismic influences. The program is based on the medium model, offered by Iwan (1967) and Mroz (1967), which is often called the IM model for short. As it is shown in the fig. 18.2, the model supposes the simulating of nonlinear curves strain-deformation, using a number of n mechanical elements, which have different stiffness kj and sliding resistance R_j, where $R_1 < R_2 < ... < R_n$. Initially the residual stresses in all elements are equal to zero. At monotonically increasing load the element j deforms until the transverse strain τ reaches R_j. After that the element j keeps positive residual stress, which is equal to R_j.

The equation, describes dynamics of soil medium, is solved by the method of central differences.

Calculation of Nonlinear Absorptive Ground Medium Vibrations Using Multiple Reflected Waves' Tool of Seismic Microzonation

Let's suppose that we have the seismic wave, which falls on the soil thickness surface. Let's assume that soil thickness is nonlinear absorptive unbounded medium with the density (and S-wave propagation velocity vs. At small deformations the value of shear modulus G will be maximum for the given soils:

$$G = G_{max} = \rho v_S^2$$

(14)

At the deformation increase the value G remains constant at first but at reaching some value (which is definite for each material or soil) the value G considerably changes, i.e. the soil begins to display its nonlinear properties. At the continued deformation increase the growth of stresses decelerates and then can remain unchanged until material destruction or hardening, i.e. until structural condition change.

As the main soil index, which characterizes its type and behavior at intensive loads, the value of plasticity PI was chosen. The parameters, which are necessary for calculations, are determined on basis of empirical ratios (Ishibashi, Zhang, 1993):

$$k(\gamma, PI) = 0.5 \left\{ 1 + \tanh \left[\ln \frac{0.000102 + n(PI)}{\gamma} \right]^{0.492} \right\}$$

(15)

Where n (PI) = {0.0 for PI = 0, $3.37 \cdot 10 - 6PI$ 1.404 for $0 < PI \le 15$, $7.0 \cdot 10 - 7$ PI 1.976 for $15 < PI \le 70$, $2.7 \cdot 10 - 5PI$ 1.115 for PI > 70;

$$d = 0.272 \left\{ 1 - \tanh \left[\ln \left(\frac{0.000556}{\gamma} \right)^{0.4} \right] \right\} e^{-0.0145PI^{1.3}}.$$

(16)

Then the change of shear modulus is determined on basis of the ratio:

$$\frac{G}{G_{max}} = k(\gamma, PI)(\sigma)^d,$$

(17)

where G is the current shear modulus, is normal stress.

Seismic energy absorption is calculated by the formula

$$\xi = 0.333 \frac{1 + \exp\left(-0.0145PI^{1.3}\right)}{2} \left[0.586 \left(\frac{G}{G_{max}} \right)^2 - 1.547 \frac{G}{G_{max}} + 1 \right]$$

(18)

On basis of the given ratios and introduced by us ratios for determination of necessary indices (normal stress, deformation etc), nonlinear version of the program ZOND was worked out. From the database of strong motions AGESAS, which was formed by us (Zaalishvili et al., 2000), the accelerogram, which was recorded on rocks in Japan, with the characteristics (magnitude, epicentral distance, spectral features etc.) similar to the territory of Tbilisi city, was chosen as the accelerogram, given into the bedrock.

The analysis of the results of linear and nonlinear calculations models of definite areas of Tbilisi city territory confirms the adequacy of calculations to the physical phenomena, which were obtained in soils at intensive loads (fig. 13) (Zaalishvili, 2009). With the increase of seismic influence intensity the nonlinearity display increases. Absorption grows simultaneously. Hence the resulting motion at quite high influence levels can be lower than the initial level. It corresponds to the fact, which is known on the results of analysis of strong earthquake consequences, which happened in recent years (for example, Northridge earthquake, 1994).

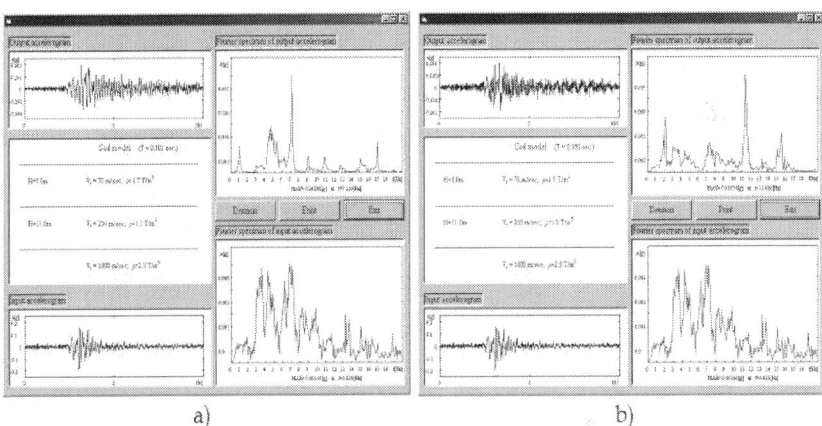

a)   b)

Figure 14: Results of calculations using multiple reflected waves' tool in linear (a) and nonlinear (b) cases.

Calculation of Nonlinear Soil Response Using Fem Tool of Seismic Microzonation

The problem of the determination of soil massif response on dynamic influence with taking soil nonlinear properties into account can be solved by usage of finite element method (FEM) in the following way (Zaalishvili, 2009).

Soil medium is represented in the form of two-dimensional massif, which is approximate by triangular finite elements. The net, which consists of triangular elements, allows to describe quite accurately any relief form and form of the layer structure of soil massif with its physics-mechanical parameters. Within finite elemet the soil is homogeneous with inherent to it characteristics, which vary in time depending on influence intensity. Earthquake accelerogram of horizontal or vertical direction, which is applied, as a rule, to the foundation of soil massif, is used as the influence. Soil is in the conditions of plane deformation and is considered as an orthotropic medium. Axes of the orthotropy coincide with the directions of main strains.

The problem of nonlinear dynamics of soil massif is solved by means of the consecutive determination of mode of deflection of the system on the previous step. The system is linear-elastic on each step.

Instrumental-Calculational Method of Seismic Microzonation

In recent years a new «instrumental-calculational» method of SMZ (per se simultaneously having the features of both instrumental and calculational method) which includes tool of «instrumental-calculation analogies» has been developed in Russia in recent years (Zaalishvili, 2006). Its usage is based on direct usage of modern databases of strong motions.

As a basis at realization of tool instrumental database of strong movements, registered in definite soil conditions, is used. As a result of given database with the help of numerical calculations it is possible more or less safety to forecast behavior of these or those soils (or their combination) for strong (weak) earthquakes with typical characteristics for the investigated territory (magnitude, epicentral distance, focus depth etc.).

Relief Influence on the Earthquake Intensity in Smz Problems

Morphological and morphometric features of relief meso- and macroforms influences on seismic intensity increment.

On basis of the analysis of numerous macroseismic observations the consequences of strong earthquakes, which took place on the territory of the former USSR, S.V.Puchkov and D.V.Garagozov offered the empirical formula for the intensity increment calculation (ΔI) depending on relief feature (Puchkov, Garagozov, 1973):

$$\Delta I = 3,3 \lg\left(W_{\mathrm{gr}}/W_{\mathrm{et}}\right) + 3,3 lg\left(W_{\mathrm{top}}/W_{fnd}\right)$$

(19)

where W_{gr}, W_{et} are the accelerations of vibratory motion on soil and etalon; W_{top}, W_{fnd} are the accelerations on the top of mountain construction and its foundation.

It was determined as a result of the instrumental and theoretical investigations that for the microrelief the increment of seismic intensity increases from the foundation of mountain-shaped feature to its top and can reach approximately 1.8 degree. For the locality mesorelilef the tendency of the increase of seismic vibration intensity from foundation to the top remains. The increment of seismic intensity for the relief mesoforms is about 0.3 degree. It was shown that weak hilly relief, with the inclinations less than 10°, does not influence on the seismic vibrations intensity.

The investigations of S.V.Puchkov and D.V.Garagozov (Puchkov, Garagozov, 1973) showed that at vibrations of mountain range, composed by volcanic tuf, the amplitude of seismic vibrations in S-waves increases on the height 15 m in 1.46 times in comparison with the foundation. For the massif, composed by loamy sand and loams on the same height marks the vibrational amplitude increased in 1.8 times for p-waves and in 3.2 times for S-waves.

Slope steepness considerably influences on the increment of seismic intensity. The increase of slope steepness, composed by incoherent gravel-pebble and sabulous-loamy grounds is conductive to the sharp worsening of engineering-geological and seismic conditions of the territory. So, for example, it is determined that slope steepness more than 19°–15° (for dry sandy-argillaceous and gravel-pebble differences) produces the intensity increment up to 1 degree and at variation of slope steepness from 10° to 40° the amplitudes of seismic vibrations increase approximately in 2.5 times.

It is known that the increase of slope steepness from 40° to 80° produces the increment of seismic intensity equal to 1.5 degree (Zaalishvili, Gogmachadze, 1989).

The correlation analysis of the dependence of seismic intensity increment on true altitude, slope steepness and relief roughness showed that the main factors, which change the value of seismic intensity, are the first two indices [Puchkov, Garagozov, 1973]. It conforms well to the investigation results of V.B.Zaalishvili, who introduced the new parameter of the relief coefficient (Zaalishvili, Gogmachadze, 1989) (fig. 14).

Later the data analysis allowed to offer us (I.Gabeeva & V.Zaalishvili) the empirical formula for the possible amplification calculation K and intensity increment ΔI, which are caused by the relief (Zaalishvili, 2006):

$$K = -0.1 + 0.68 \lg R$$

(20)

where R= H is the relief coefficient; is the relief slope angle, degree; H is height, m.

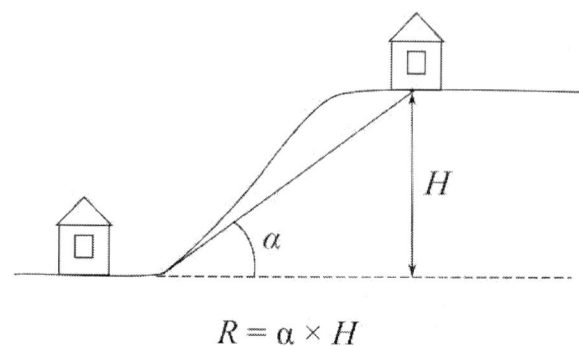

$$R = \alpha \times H$$

Figure 15: Relief coefficient RThe analysis of the experimental data shows that intensity increment can vary at that independently of the type of rocks, from 0 to 1.5 degree.

Finally, let's try to assess the amplification of vibrational amplitude, which is caused by relief, with the help of the calculation method of FEM (Zaalishvili, 2006).

The algorithm for the calculation of seismic reaction of soil thickness for the two-dimensional model was developed for this purpose (fig. 15) (Zaalishvili, 2009). The results of the executed earliear investigations were used for the program testing (Puchkov, Garagozov, 1973). Mountain structure had the form of frustum of a cone with the height 30 m and slope angle of the generatrix 30°. The element maximum size was equal to 5 m, S-wave propagation velocity was 300 m/s, the density 1800 kg/m³. The seismic influence was applied to the foundation of soil thickness in the form of instrumental accelerogram, modeling the vertically propagating SH wave.

It was determined that the vibrational amplitude considerably chances with the relief. The given dependence at that is various for the displacements, velocities and accelerations. The largest value of the amplification is observed for displacements and the maximum ratio of vibrational amplitudes, for example, in the point C to the point A, is 2.1 and for the point D – 3.2. It well satisfies the results of experimental observations where the ratio in the point C for the S-wave is equal to 2.3 and in the spectral region the maximum values are 1.8 (at T = 0.4 s) and 3.2 (at T = 0.7 s) for P- and S-waves accordingly. Spectral analysis also shows the resonance increase of vibrational amplitudes in the top part of the slope on the frequency 1.6 Hz (i.e. T=0.6 s).

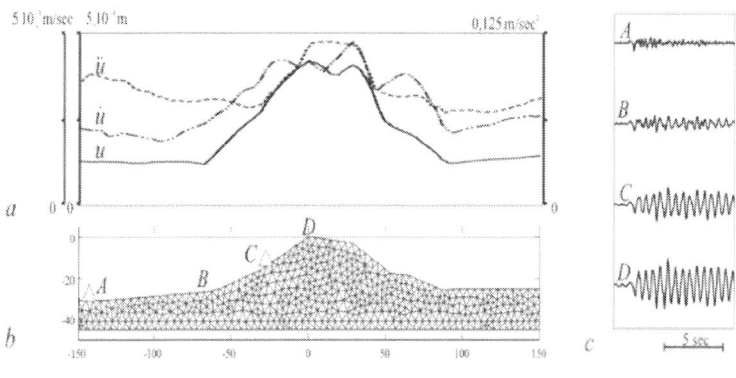

Figure 16: Final elements analysis (FEA) application example: a) Variation of amplitudes of displacement, velocity and acceleration along surface; b) calculational model; c) seismograms, calculated in points A, B, C, D.

Considerably fewer investigations are dedicated to the influence of the underground relief on the intensity. On the data of B.A.Trifonov (1979) the underground and buried topography of the rocks influences on seismic vibrations intensity, if the surface slope exceeds 0.3. At the vee couch of the rocks, which are covered by sedimentary thickness, the ratio between wave length and the sizes of vee stripping influences on seismic intensity change. Seismic intensity increment in the given case is formed by the wave interference and can be 1.5–2.0 degree (Bugaev & Kharlov, 1977; Bondarik et.al., 2007).

Thus, at the execution of SMZ works in the mountain regions or under the conditions of billowy relief, it is necessary to pay special attention to the influence of surface or underground relief on the intensity forming. It is necessary to continue the investigations in order to obtain statistically proved ratio for the calculation of intensity increment, caused by relief.

Seismic Microzonation of Vladikavkaz City

If we consider 5% DSZ map as basis for seismic microzonation so seismic intensity of 8 corresponds to etalon grounds for whole territory.

Then, maps of seismic microzonation of cities must be created. According to the above mentioned maps of detailed zoning the maps of seismic microzonation with probability 1%, 2%, 5% or 10 %, correspondingly, were made up.

Though, that definitions of the word «zoning» are similar, actually they are quite different in essence. Unlike the maps of detailed seismic zoning, which give seismic potential (Mmax) and source features, the maps of seismic microzonation give assessments of soil condition influence (sands, rocks, pebbles, clays etc., their combination; watering; relief (as underground as surface); spectral distribution of incoming wave; predominant vibration frequencies on city square etc.) on forming of future earthquake intensity. As a rule, the scale of such maps is 1:10 000, in order to have the opportunity of taking them into account at building. Maps can be more detailed (1:5000 etc.) but this makes no sense as the type and physical condition of soils in space on the territory site can change fast. The most important thing is to assess intensity of possible earthquakes on areas with typical soil conditions for city territory.

Maps of seismic microzonation can be made up for the certain territories (cities and settlements, as a rule). It is impossible to make them up in entire format because of the necessity of geological conditions knowledge on larger territories, which are mostly not built up. We often don't have such data even for the modern cities! It's practically impossible because the resources will be lost for nothing! And absurdity! In the other words there is no the microzonation map even for the territory of North Ossetia let alone the whole Northern Caucasus.

Maps of seismic microzonation do not only show the place of earthquake-proof building up, but they also show on what intensity this or that building must be calculated and designed: on 6, 7, 8 or 9 points. And sometimes even on 10 points (for very soft grounds!). And this suggests investments of different financing for the realization of antiseismic measures (thicker armature, more connections etc.). Seismic risk can considerably be reduced at building-up zones with 7, 8 and 9 point of the calculated intensity by adequate site development on the territory of city, for example, as social losses will be minimal, though buildings will be damaged in this or that extent.

In the next stage we should carry out SMZ. It should be noted that as a basis the maps of different probability of exceedance will be used and as the initial intensity, the value of which corresponds directly to the intensity of the sites, composed by average soils or characterized by average soil conditions and, therefore, the maps will be referred to the 7, 8 or 9 points (and similarly for acceleration). The zones, composed by clay soils of fluid consistency, which can be characterized by liquefaction at quite strong influences, are marked by the index 9*. Intensity calculation here supposes the usage of special approaches in the form of direct taking soil nonlinearity into account (Zaalishvili, 2000). The usage of relevant methods and techniques of SMZ will allow to obtain the correspondent maps of SMZ.

Thus for maps with probability of exceedance 1%, 2%, 5% and 10% one can obtain corresponding maps of SMZ with probability of exceedance 1%, 2%, 5% and 10%, i.e. probabilistic maps of SMZ (Fig. 16).

For each of the zoning subject the probabilistic map of the seismic microzonation with location of different calculated intensity (7, 8, 9, 9*) zones is developed (the zones, composed by clay soils of fluid consistency, which can be characterized by liquefaction at quite strong influences, are marked by the index 9*). The maps in accelerations units show the similar results.

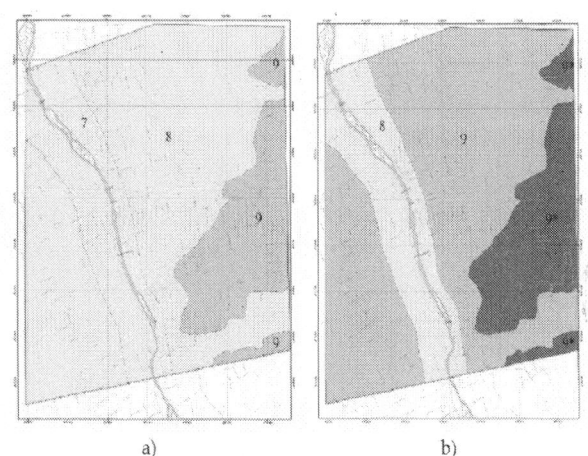

a)   b)

Figure 17: The maps of seismic intensity microzonation for probabilities of 5% (a) and 2% (b) for the central part of Vladikavkaz city territory (Zaalishvili et al., 2010).

Such maps of SMZ except of mentioned developments are also based on materials of local network of seismic observations "Vladikavkaz". Network was organized for the first time on the urbanized territory of the Northern Caucasus in July 2004. Stations are located on the sites with different typical for the city soils (clays of medium-hard and liquid consistence, gravels with filling material of less than 30% and more than 30%, and their assembly).

It must be noted that usage of the maps with high time exposition i.e. maximal magnitude (maximal intensity) for given territory (for return period of 50 years and exceedance probability 2% or 1%) physical nonlinearity of soils necessarily must be taken into account with the help of developed tools (Zaalishvili, 2009).

Unlike small-scale M 1:8 000 000 seismic hazard map of the territory of Russia (GSZ) maps of DSZ in scale 1:200 000 allow taking into account features of specific seismic sources (faults) directly. But the main thing is that such scale zoning is suitable for quite large territories. So it's seen that alignment of faults of different constituent entities of the Russian Federation of Northern Caucasus make a good sense (fig.7).

SPECIFIED SEISMIC FAULT AND DESIGN SEISMIC MOTION

Analysis and consequent account of initial accelerograms transformation will become the basis for site effect analysis at strong seismic loadings (fig. 17) (Zaalishvili et al., 2010).

Methods of such modeling are based on accordance of spectral properties of modeled and real earthquake. In a whole modeling accuracy depending on the purposes of total motion usage and what characteristics defining structural system behavior must be reproduced.

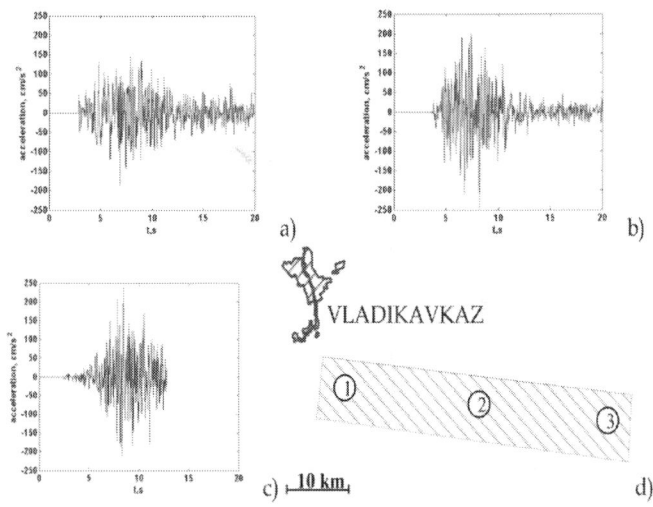

Figure 18: Synthetical accelerograms for different source locations: a – western part of fault; Sb – middle part of fault; c – eastern part of fault; d – scheme of sources of scenarios earthquakes.

Earthquake source that is a region of rupture can be considered as point source only for much larger distances than fault size. At close distances effects of finite fault size become more significant. Those phenomena are mainly connected with finite rupture velocity, which causes energy radiation of different fault parts in different times and seismic waves are interference and causes directivity effects (Beresnev & Atkinson, 1997, 1998).

Let's compare amplitude spectra of obtained design accelerograms with spectrum of real earthquake from considered fault. Data analysis (fig. 18 and fig. 19) shows that spectra of calculated and real earthquakes in a whole are similar in their main parameters.

It must be noted that spectrum of vertical component of real earthquake is closer to design spectra. The last fact is quite obvious and is explained by proximity to earthquake source. Indeed, close earthquakes in general are characterized by predomination of vertical component. Record of TEA station (located in theater) was selected due to its location on dense gravel and has a minimal distortions caused by soil conditions.

Analysis of spectrum of weak earthquake shows that peaks are observed on 1.3 and 5.6 Hz (Fig. 18). In spectra of synthesize accelerograms mentioned amplitudes are also observed. At the same time medium response on strong earthquake, undoubtedly, differ from weak earthquake response (Fig. 19)(Zaalishvili, 2000).

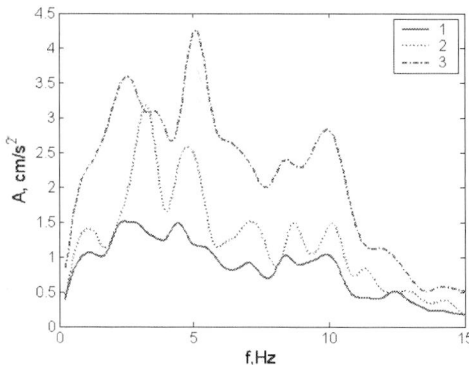

Figure 19: Spectra of design accelerograms at different source locations of earthquake M = 7,1 : 1 – western part of fault; 2 – middle part of fault; 3 – eastern part of fault.

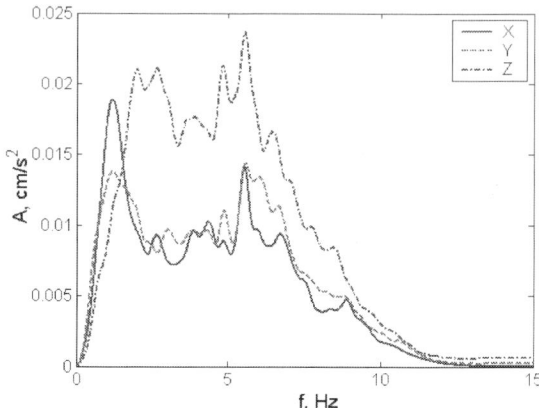

Figure 20: Spectra of accelerograms of weak earthquake with epicenter in the zone of Vladikavkaz fault. (25.08.2005 10:25 GMT, H = 8 km M= 2.5).

Usage of maps of detailed seismic zoning in units of accelerations at seismic microzonation level is possible only for calculation method giving results in units of accelerations. Today traditional instrumental method of seismic microzonation does not allow obtaining intensity increments in accelerations due to traditional orientation on macroseismic intensity indexes. The exclusion is the case of investigation of strong earthquakes accelerations when instrumental records are obtained (in presence of accelerometer) (Zaalishvili, 2000). At the same time investigations are conducted and the problem supposed to be solved.

On the other hand in recent years a new instrumental-calculation method was developed (Zaalishvili, 2006). New method is based on selection from database (including about 5000 earthquake records) soil conditions which are the most appropriate to real soil conditions of the investigated site. Then the selection of seismic records with certain parameters or their intervals follows (magnitude, epicentral distance, and source depth). Then maximal amplitudes are recalculated for given epicentral distances. Absorption coefficient can be calculated by attenuation model for given region.

Thus, a new complex method of seismic hazard assessment providing probability maps of seismic microzonation, which are the basis of earthquake-proof construction, is introduced. Undoubtedly such approach significantly increases physical validity of final results.

Considered procedures on the level of possible seismic sources zones exploration, maps of detailed seismic zoning and seismic microzonation may differ from described above. So paleoseismological investigations like «trenching» (Rogozhin, 2007), which allow determining more reasonable the recurrence and other features of seismic events realization are also possible when it is necessary.

Today, we have conditions for detailed seismic zoning maps development like the above mentioned but for all the territory of the Northern Caucasus on basis of the modern achievements of engineering seismology. It will give us a possibility to develop probabilistic maps of seismic microzonation with the help of powerful nonexplosive sources, methods taking into account physical soils nonlinearity (Zaalishvili, 2009).

Thus algorithm of seismic hazard assessment of the territory taking into account multiple factors forming seismic intensity was considered. Forms of typical seismic loadings for firm soils are given, which will be changed from site to site in dependence of differences in ground conditions (engineering-geological, geomorphological and gidrogeological conditions)

REFERENCES

1. F. F. Aptikaev, et al.1986Methodological recommendations on detailed seismic zoning. Questions of engineering seismology. 27Moscow, 1986. 184212in Russian)

2. A. R. Arakelyan, V. B. Zaalishvili, V. D. Makiev, D. A. Melkov, 2008To the question of seismic zonation of the territory of the Republic of North Ossetia-Alania / Procs. of Ist International conference "Dangerous natural and man-caused processes on the mountaneous and foothill territories of Northern Caucasus", Vladikavkaz September 20-22, 2007. Vladilavkaz: VSC RAS and RNO-A, 2008, 263278in Russian).

3. J. P. Bardet, T. Tobita, N. E. R. A. A. computer, for. program, Earthquake. Nonlinear, Response. site, of. Analyses, soil. layered, deposits, Univ. of Southern California, Los Angeles, 2001p.

4. J. P. Bardet, K. Ichii, C. H. Lin, 2000EERA, A Computer Program for Equivalent Linear Earthquake Site Response Analysis of Layered Soils Deposits. University of Southern California, Los Angeles

5. P. Bazzurro, C. A. Cornell, 1999Disaggregation of Seismic Hazard, Bull. Seism. Soc. Am. 89, 2, 501520

6. B. Bender, D. M. Perkins, 1987SEISRISK III: A Computer Program for Seismic Hazard Estimation. US Geological Survey Bulletin 1772, 48p.

7. I. A. Beresnev, G. M. Atkinson, 1997Modeling finite fault radiation from ωn spectrum. Bull. Seism. Soc. Am., 876784

8. I. A. Beresnev, G. M. Atkinson, 1998FINSIM- a FORTRAN program for simulating stochastic acceleration time histories from finite faults. Seismological Research letters. 691

9. G. K. Bondarik, V. V. Pendin, L. A. Yarg, 2007Engineering geodynamics. Moscow: "Universitet". 440 p. (in Russian)

10. G. Bonnet, J. F. Heitz, Non-linear seismic response of a soft layer // Proc. of the 10th European Conf. on Earthquake Eng.Vienna. 19941361364

11. E. G. Bugaev, E. M. Kharlov, 1977Features of canion sides vibrations. Seismic microzonation. Moscow: "Nauka". 9198in Russian)

12. E. I. Byus, 1955aSeismic conditions of Transcaucasus. Part I. Tbilisi: Academy of Sciences of USSR, 1948 (in Russian).

13. E. I. Byus, 1955bSeismic conditions of Transcaucasus. Part II. Tbilisi: Academy of Sciences of USSR, 1952 (in Russian).

14. E. I. Byus, 1955cSeismic conditions of Transcaucasus. Part III. Tbilisi: Academy of Sciences of USSR, 1955 (in Russian).

15. T. Chelidze, Z. Javakhishvili, 2003Natural and technological hazards of territory of Georgia: implications to disaster management. Journal of Georgian Geophysical Society. Issue (A) Solid Earth, 8318

16. C. A. Cornell, 1968Engineering risk in seismic analysis. Bull. Seism. Soc. Am. 54 1968, 5831606

17. C. A. Cornell, Engineering risk in seismic analysis. Bull. Seism. Soc. Am. 5419681968583

18. I. Gamkrelidze, T. Giorgobiani, S. Kuloshvili, G. Lobjanidze, G. Shengelaia, 1998Active Deep Faults Map and the Catalogue for the Territory of Georgia // Bulletin of the Georgian Academy of Sciences, 157, 18085

19. G. P. Gorshkov, 1984Regional seismotectonics of the territory of south of USSR. Moscow: "Nauka", 1984. 272 p. (in Russian)

20. I. Ishibashi, X. Zhang, 1993Unified dynamic shear moduli and damping ratios of sand and clay," Soils and Foundations, 331182191

21. Z. Javakhishvili, O. Varazanashvili, N. Butikashvili, 1998Interpretation of the Macroseismic field of Georgia. Journal of Georgian Geophysical Society. Issue (A) Solid Earth, 38588

22. K. Kanai, Relation between the nature of surface layer and the amplitudes of earthguake motions // Bul. Earthquake Res. Inst. 30Tokyo Univ. 19523137

23. A. B. Maksimov, 1969Methodology of microzonation on the basis of detailed investigation of seismic properties of soils. Kandidate of phys.-math. sciences dissertation abstract. Moscow, 1969(in Russian)

24. S. Mc Clusky, S. Balassanian, C. Barku, et al.2000Global Position System constraints on plate kinematics and dynamics of the Mediterranean and Caucasus // J. Geophys. Res. 2000, 105B35569555719

25. R. Mc Guire, 1976FORTRAN computer program for seismic risk analysis, US Geological Survey, open file report, 7667

26. R. Mc Guire, 1995Probabilistic Seismic hazard analysis and design earthquakes: closing the loop. 83512751284

27. S. V. Medvedev, 1947On the question of taking into account seismic activity of region at construction. Procs. of seismological institute of AS USSR. 119in Russian)

28. S. V. Medvedev, 1962Engineering seismology. Moscow: Gosstroyizdat, 1962. 284 p. (in Russian)

29. I. V. Mushketov, 1889Venensk earthquake of May 28 (June 9) 1887. Procs of geological comm. 1889. 101in Russian)

30. I. V. Mushketov, geology. Physical, Petersburg. St, 1891Part. 1. 709 p. (in Russian)

31. R. Musson, 1999Probapilistic seismic hazard maps for the North Balkan region. 1999. Annali di Geofisica. 42611091124

32. Y. A. Nakamura, for. Method, Characteristics. Dynamic, of. Estimation, using. Subsurface, on. Microtremor, Ground. the, Q. Surface, QR of RTRI, 3011989

33. Yu. V. Nechaev, G. I. Reisner, E. A. Rogozhin, et al.1998Geological-geophysical and seismological criteria of potencial seismicity of Western Caspian // Exploration and protection of subsurface resources. 1998, 21316in Russian).

34. S. A. Nesmeyanov, 2004Engineering geotectonics. Moscow: Nauka, 2004. 780 p. (in Russian)

35. S. A. Nesmeyanov, I. I. Barkhatov, 1978Newest seismogenic structures of Western Gissaro-Alay. Moscow: Nauka, 1978. 120 p. (in Russian)

36. New Catalogue of strong Earthquakes in the USSR from Ancient times through1977NOAA, USA, 1521

37. A. V. Nikolaev, 1965Seismic properties of grounds. Moscow: Nauka, 1965. 184 p. (in Russian)

38. A. V. Nikolaev, 1987Problems of nonlinear seismics. Moscow: Nauka, 1987. 520in Russian)

39. P. Smit, V. Arzmanian, Z. Javakhishvili, S. Arefiev, D. Mayer-Rosa, S. Balassanian, T. Chelidze, 2000The Digital Accelerograph Network in the Caucasus. In: "Earthquake Hazard and Seismic Risk Reduction". Kluwer Academic Publishers, 109118

40. Paleoseismology of Great Caucasus1979Moscow: Nauka, 1979, 188 p. (in Russian)

41. A. Poceski, The Ground effects of the Scopje July 26, 1963 Earthquake, BSSA. 1969591122

42. S. V. Puchkov, D. Garagozov, 1973Investigation of hilly relief of region on intensity of seismic vibrations during earthquakes. Problems of engineering seismology. 15Moscow: Nauka, 1973. 9093in Russian)

43. E. Rantsman, Ya, 1979Places of eartquakes and morphostructure of mountainous countries. Moscow: Nauka, 1979. 171 p. (in Russian)

44. Recommendations on seismic microzonation (SMR-73).Influence of grounds on intensity of seismic vibrations. (1974Moscow: Stroyizdat, 1974. 65 p. (in Russian)

45. Recommendations on seismic microzonation at engineering survey for construction1985Moscow: Gosstroy USSR, 1985. 72 p. (in Russian)

46. G. I. Reisner, L. I. Ioganson, Complex typification of earth crust as basis for fundamental and applied tasks solution. Article 1 and 2. Bull. MOIP, 1997Geology dept., 723513in Russian).

47. L. Reiter, hazard. Earthquake, New. analysis, Columbia. York, Press. Univ, 1991p.

48. Yu. V. Riznichenko, 1966Calculation of points of Earth surface shaking from earthquake in surrounding area. Bull. of AS of USSR. Physics of the Earth. 1966. 5. 1632in Russian)

49. E. A. Rogozhin, 1997Geodynamics and seismotectonics. in Problems of evolution of tectonosphere. Moscow, 1997. 8492in Russian)

50. E. A. Rogozhin, G. I. Reisner, L. I. Ioganson, 2001Assessment of seismic potencial of Big Caucasus and Apennines by independent methods // Modern mathematical and geological models in applied geophysics tasks: selected scientific works. Moscow: UIPE RAS, 2001, 279300in Russian).

51. E. A. Rogozhin, 2007PSS zones and their characteristics for the territory of the Republic of North Ossenia-Alania. Procs. Of VI international conference "Innovative technologies for sustainable development of mountainous territories" May 28-30 2007. Vladikavkaz: "Terek", 2007. 283in Russian)

52. E. A. Rogozhin, A. N. Ovsyuchenko, A. V. Marakhanov, S. S. Novikov, B. V. Dzeranov, D. A. Melkov, 2008Research report "Investigations of marks of possible occurrence of seismic activity in the zone of Vladikavkaz fault". Vladikavkaz, 2008, 1book 8, 33 p., (in Russian).

53. P. B. Schnabel, J. Lysmer, H. B. Seed, 1972SHAKE: A Computer Program for Earthquake Response Analysis of Horizontally Layered Sites", Report No. UCB/EERC-72/12, Earthquake Engineering Research Center, University of California, Berkeley, December, 102p.

54. Seismic zoning of USSR terrytory.Methodological basics and regional description of the map of 1978Moscow: Nauka, 1980. 308 p. (in Russian)

55. V. V. Shteinberg, 1964Analysis of grounds vibrations from close earthquakes /Procs. of IPE RAS 33Moscow, 1964. 1124in Russian)

56. V. V. Shteinberg, 1965Influence of layer on amplitude-frequency spectrum of vibrations on the surface. Seismic microzonation. / Questions of engineering seismology. Moscow: Nauka, 1965. 3435in Russian)

57. V. V. Shteinberg, 1967Investigation of spectra of close earthquakes for prognosis of seismic impact.- Vibrations of earth dams / Questions of engineering seismology Moscow: Nauka, 1967. 123150in Russian)

58. SP2011Construction works in seismic regions. Actualized version of SNiP II-781Minregion of Russia.- M. : «TsPP Ltd», 2011.- 167 p.

59. Mihailov. V. Stoykovic, Some results of the investigations in the seismic microzoning of Banja Luka // Proc. 5th World Conf. on Earthquake Eng. 1Rome, 197317031708

60. V. G. Trifonov, 1999Neotectonics of Eurasia. Moscow: Nauchniy Mir, 1999, 252 p. (in Russian)

61. V. I. Ulomov, 1995About main thesis and technical recommendations on creation of new map of seismic zoning of the territory of Russian Federation. Seismicity and seismic zoning of Northern Eurasia. Moscow: UIPE RAS, 1995. 2626in Russian)

62. V. I. Ulomov, L. S. Shumilina, V. G. Trifonov, et al.1999Seismic Hazard of Northern Eurasia // Annali di Geofisica, 42610231038

63. V. B. Zaalishvili, 1986Seismic microzonation on the data of artificial vibrations of ground thickness. Candidate of phys.-math. sciences dissertation abstract. Tbilisi, 1986a. (in Russian)

64. V. B. Zaalishvili, S. A. Gogmachadze, 1989Influence of relief on wave field of pulse and vibrational sources. Investigation of fields of pulse and vibrational sources for the means of seismic microzonation: Report of ISMIS AS GSSR. Tbilisi, 1989. 2540in Russian)

65. V. B. Zaalishvili, 1996Seismic microzonation on the basis of nonlinear properties of grounds by means of artificial sources. Doctor of phys.-math. sciences dissertation abstract. Moscow: MSU, 1996. (in Russian)

66. V. Zaalishvili, M. Otinashvili, Z. Dzhavrishvili, 2000Seismic hazard assessment for big cities in Georgia using the modern concept of seismic microzonation with consideration soil nonlinearity. INTAS/Georgia/970870Periodic report. 2000. 170p.

67. V. B. Zaalishvili, 2000Physical bases of seismic microzonation. Moscow: UIPE RAS, 2000. 367 p. (in Russian).

68. V. B. Zaalishvili, 2006Basics of seismic microzonation. VSC RAS&RNO-A. Vladikavkaz, 2006. 242 p. (in Russian)

69. V. B. Zaalishvili, 2009Seismic microzonation of urban territories, settlements and large building sites. Moscow: Nauka, 2009, 350 p. (in Russian).

70. V. B. Zaalishvili, D. A. Melkov, O. G. Burdzieva, 2010Determination of seismic impact on the basis of specific engineering-seismological situation of region // "Earthquake engineering. Buildings safety", 2010 13539in Russian).

71. V. B. Zaalishvili, E. A. Rogojin, 2011Assessment of Seismic Hazard of Territory on Basis of Modern Methods of Detailed Zoning and Seismic Microzonationю The Open Construction and Building Technology Journal, 2011, 53040

CITATION

V. B. Zaalishvili (2012). Assessment of Seismic Hazard of Territory, Earthquake Engineering, Prof. Halil Sezen (Ed.), ISBN: 978-953-51-0694-4, InTech, DOI: 10.5772/48324.

Index